Pneumatics

FUNDAMENTALS OF PNEUMATIC CONTROL ENGINEERING

Text book

A text book from

2nd edition

© Copyright by FESTO DIDACTIC D-7300 Esslingen 1978
All rights reserved.
Printed in W. Germany
Authors: J. P. Hasebrink, R. Kobler.

ISBN 3-8127-0851-5

Preface to the 1st edition

In the course of the tumultuous development in the field of automation, the field of control engineering has also achieved a position of increasing importance. The problems to be solved have become more demanding and more complex, and consequently the requirements imposed on an individual control system have also increased.

Reliability, dependability in the operating sequence, as well as speed of the installation and hence also of the control, are points which can at the present time no longer be neglected. New, complicated, and versatile elements in the various technical groups of equipment have been developed and placed on the market, whereas the education of the people who are to, or have to, work with these technologies has in many cases not progressed at all. Indeed, in many institutions there is not even the opportunity of receiving education or further training in the specialized field of control engineering. The object of this and succeeding brochures and seminars is to at least partially fill this gap, and one branch of control engineering – the field of cylinder controls – will be dealt with in detail.

This book, the first part of Control Engineering, covers as the title indicates the fundamentals of pneumatic and electropneumatic controls. The aim is to get the reader and seminar participant to the point where he is able to solve simpler types of control problems and develop circuit diagrams based on both types of engineering principles dealt with here.

This path leads from an introduction to the general fundamentals of control engineering on to the technicalities of individual items of equipment; symbolism, definition of terms, and basic controls are dealt with first.

An indication is provided of the different ways of handling problems and constructing circuit diagrams, although in the scope of the present work only the simpler methods can be dealt with. To delve deeper into the methodology of constructing circuit diagrams calls for a knowledge of switching algebra and logic in addition to elementary knowledge provided here. The second part of Control Engineering and Seminar E 2 are concerned with these problems.

For the sake of clarity, this book has been divided into four parts.

Part 1 deals with the general fundamentals of control engineering. In Parts 2 and 3, the characteristic features of pneumatics and electropneumatics are discussed, and electromechanical controls are presented in the section on electropneumatics.

To supplement the examples already given in Parts 2 and 3, Part 4 contains further interesting control problems encountered in practice. The reader should also be encouraged to collect and file examples.

As regards the structure of the brochure itself, the material has been prepared from a didactic viewpoint. Problems, examples, and exercises are designed to ensure learning success. It has been attempted to observe existing standards and regulations and to some extent to include them.

Finally, it should be mentioned that the reader of this text and the seminar participant is expected and assumed to have a certain background knowledge. The required knowledge concerns the elements and devices needed in pneumatic and electrical controls, their characteristics and behaviour, and also a knowledge of the characteristics of the energy carriers air and electricity.

The authors

Preface to the 2nd edition

Since the 1st edition, pneumatics has continued to develop. The areas of application for pneumatic controls have expanded, the completed controls have become more elaborate. At the same time, even more is expected of controls.

The 2nd edition has been brought into line with this situation.

Finally, we wish to thank all who have contributed to the revision and expansion by their suggestions and comments.

The authors

Contents

Page

Part 1: Fundamentals of control engineering

1.1	Introduction	9
1.2	Introduction to the subjects of control and automatic control Definition of terms	9
1.3	Various terms in automatic control and control engineering	11
1.3.1	Actuating path and direction of action	12
1.3.1.1	Constituent sections of the actuating path	12
1.3.1.2	Variables and ranges of variables in the actuating path	12
1.3.2	Signals	13
1.3.3	Signal flow diagram	15
1.3.4	Summary	16
1.4	Breakdown of the control chain	17
1.5	Types of energy for operative part and control part – summary and definition	20
1.5.1	Working media	21
1.5.2	Control media	22
1.6	Differentiating characteristics of controls	23
1.6.1	According to the control energy used	23
1.6.2	According to the mode of operation with respect to signal processing	23
1.6.3	According to the type of operating sequence	24
1.6.4	Pilot control	24
1.6.5	Memory control	25
1.6.6	Program controls	25
1.6.6.1	Time-schedule control	26
1.6.6.2	Coordinated motion control	26
1.6.6.3	Sequence control	26
1.6.7	Comparison and definitions of program control systems	27
1.7	Means of representing motion sequences and switching conditions	27
1.7.1	Writing down in chronological sequence	28
1.7.2	Tabular form	28
1.7.3	Vector diagram	28
1.7.4	Abbreviated notation	28
1.7.5	Graphical representation in diagram form	28
1.7.5.1	Motion diagrams	29
1.7.5.2	Control diagram	30
1.7.6	Symbols and representation standards	31
1.8	Working out a control problem	37
1.8.1	Problem definition, determination of conditions	37
1.8.2	Working energy, working elements	38
1.8.3	Positional sketch	38
1.8.4	Determination of the sequence of operations	38
1.8.5	Choice of type of control	38
1.8.6	Choice of control energy	38
1.8.7	Circuit diagram	38
1.9	Exercise	38
1.9.1	Procedure for working out the exercise	39
1.9.1.1	Definition of problem and conditions	39
1.9.1.2	Selection of working energy and dimensioning of working elements	39
1.9.1.3	Positional sketch	40
1.9.1.4	Determination of sequence of operations	40
1.9.1.5	Selection of type of control	41
1.9.1.6	Control energy	41

Part 2: Pneumatic Controls

		Page
2.1	Indroduction	42
2.2	List of symbols used	43
2.3	Diagrammatic representation of pneumatic circuit diagrams	57
2.3.1	Building-up the circuit diagram	57
2.3.2	Designating the elements	58
2.3.2.1	Designation using digits	58
2.3.2.2	Designation using letters	59
2.3.3	Representation of devices	60
2.3.4	Graphic symbols	61
2.3.5	Pipelines	61
2.3.6	Pipeline designations	61
2.3.7	Additional information in the circuit diagram	61
2.3.8	Modified arrangements	62
2.3.9	Summary	62
2.4	Basic pneumatic circuits	62
2.4.1	Exercises on basic controls	63
2.4.2	Basic circuits using directional control valves	64
2.4.2.1	Control of a single-acting cylinder	64
2.4.2.2	Control of a double-acting cylinder	64
2.4.2.3	Indirect control of a single-acting cylinder	64
2.4.2.4	Indirect control of a double-acting cylinder	65
2.4.2.5	Indirect control of a single-acting cylinder using a holding control	65
2.4.2.6	Automatic return control of a double-acting cylinder using a limit switch	67
2.4.2.7	Continuous reciprocation of a double-acting cylinder with means of switching off	68
2.4.2.8	Stopping and fixing a double-acting cylinder in intermediate positions	68
2.4.2.9	Summary	69
2.4.3	Circuits for speed regulation on cylinders	70
2.4.3.1	Speed regulation on a single-acting cylinder	71
2.4.3.2	Speed regulation on a double-acting cylinder	72
2.4.4	Circuits with shuttle and two-pressure valves	74
2.4.4.1	Shuttle valve	74
2.4.4.1.1	Controlling a single-acting cylinder from two different points	74
2.4.4.2	Two-pressure valve	75
2.4.4.2.1	Control of a double-acting cylinder by means of two valves and one two-pressure valve	76
2.4.5	Pressure operated controls	77
2.4.5.1	Pressure controlled reversal with mechanical end-position checking using limit switches	77
2.4.5.2	Pressure controlled reversal without mechanical checking of the end position	78
2.4.6	Circuits with time behaviour	78
2.4.6.1	Time circuits for defined time-dependent reversal	79
2.4.6.2	Time circuits for pulse shaping	80
2.4.6.3	Exercise 2.4.1.14	80
2.4.7	Basic circuits with contactless signal transmitters	81
2.4.8	Various basic circuits	84
2.4.8.1	Alternating controls	84
2.4.8.1.1	Alternating circuit with latching by means of limit switches	85
2.4.8.1.2	Alternating circuit with latching by means of impulse valve and shuttle valve	85
2.4.8.1.3	Alternating circuit with latching by means of impulse valve and pressurized 3-way valve with spring return	86

© by FESTO DIDACTIC

		Page
2.4.8.1.4	Alternating circuit with latching by means of impulse valve and two-pressure valve	87
2.4.8.1.5	Valves with alternating behaviour	87
2.4.8.2	Circuit for reciprocating movements of a cylinder without limit switch	88
2.4.9	Circuits for signal suppression and signal elimination	89
2.4.9.1	Circuits for signal suppression	89
2.4.9.2	Circuits for signal elimination	89
2.4.9.3	Summary	91
2.5	Methods for constructing a circuit diagram	92
2.6	Constructing the circuit diagram for coordinated motion controls	93
2.6.1	Exercise 6.1 Package transfer	94
2.6.1.1	Circuit design for exercise 6.1	94
2.6.2	Exercise 6.2 Riveter	97
2.6.2.1	Construction of circuit diagram for exercise 6.2 with signal cut-out by means of idle return rollers	97
2.6.2.2	Construction of circuit diagram for exercise 6.2 with signal cut-out by means of reversing valve	99
2.6.2.3	Design of a cascade	102
2.6.2.4	Design of a shift register	105
2.6.2.5	Method for designing coordinated motion controls with cascade or shift register	107
2.6.2.6	Procedure for composing a circuit diagram by the Block Method	109
2.6.3	Exercise 6.3 Stamping appliance	111
2.6.3.1	Construction of circuit diagram for exercise 6.3 with signal switch-off via idle return rollers	112
2.6.3.2	Circuit diagram for exercise 6.3 with signal switching via reversing valve Design by the Block Method	114
2.6.4	Exercise 6.4 Bending fixture	120
2.6.4.1	Circuit for exercise 6.4 with signal cut-out via idle return rollers	121
2.6.4.2	Circuit for exercise 6.4 based on the Cascade Method	122
2.6.5	Exercise 6.5 Shearing unit	124
2.6.5.1	Circuit for exercise 6.5 based on the Cascade Method	125
2.6.6	Exercise 6.6 Pressing fixture	126
2.6.6.1	Circuit for exercise 6.6 with idle return rollers	127
2.6.6.2	Circuit for exercise 6.6 with cascade and shift register	128
2.6.7	Exercise 6.7 Drilling unit	132
2.6.7.1	Solution to exercise 6.7	133
2.7	Constructing circuit diagrams for pilot controls	135
2.7.1	Exercise 7.1 Test station for cans	135
2.7.1.1	Circuit design for problem 7.1	135
2.7.2	Exercise 7.2 Test station	137
2.7.2.1	Circuit design for exercise 7.2	137
2.8	Constructing the circuit diagram for memory controls	139
2.8.1	Exercise 8.1 Pneumatic switching section	139
2.8.1.1	Solution for exercise 8.1	140
2.9	Constructing the circuit diagram for time-schedule controls	141

Part 3: Control Problems, Examples and Solutions

3.1	Clamping fixture	143
3.1.1	Circuit for exercise 3.1	143
3.2	Door control	144
3.2.1	Circuit for exercise 3.2	144
3.3	Bending fixture for spectacle frames	145
3.3.1	Circuit for exercise 3.3	146

			Page
3.4	Elevator control		147
3.4.1	Circuit for exercise 3.4		148
3.5	Transporting of section material		149
3.5.1	Circuit for exercise 3.5		150

Part 4: Appendix
4.1	Examples of circuit diagram drawings	151

© by FESTO DIDACTIC

Part 1: Fundamentals of control engineering

1.1 Introduction

Little need be said about the significance of automatic control and control engineering for industrialized society. Without these fields, the present-day advanced stage of technology would be inconceivable. Control systems are required in all branches of engineering. A continuous and often tumultuous development in these fields was, and will also be the future, the direct result of this necessity.

During the course of further development producing both entirely new systems and equipment as well as a continuous improvement and expansion of existing systems and elements, it was also necessary to continually extend, modify, or even completely revise existing regulations and standards.

In order to allow collaboration on a wider scale, a uniform language is essential; i.e. precise definitions of terms must be drawn up and universally valid fundamental principles must be worked out.

In keeping with this, it will be attempted in Part 1 to provide an outline of the more important terms which are valid and currently used in automatic control and control engineering at the present time. It is necessary to include the basic terms used in automatic control engineering because the two fields do in fact have much in common, in particular terms and designations, and these fields increasingly overlap each other. Consequently however, it is also necessary to define an exact demarcation – this is provided in Section 1.2.

The fundamental principles of control engineering dealt with here apply to the subject as a whole and are therefore applicable regardless of the control energy used or of the type of equipment used in the control.

1.2 Introduction to the subjects of control and automatic control Definition of terms

The standard which is used for terms and designations in automatic control and control engineering is DIN 19 226 ("Regelungstechnik und Steuerungstechnik, Begriffe und Benennungen" – "Automatic Control Engineering and Control Engineering, Terms and Designations"). The available version is dated May 1968 and is one of the reference sources for the following definition.

Before going into this however, some other definitions for the term control, taken from publications and general language usage, should be mentioned briefly and without comment.

Control: "Appliance for influencing larger energies by smaller energies"

"The whole of the components with which the performance of a machine or the operation of equipment is changed, usually automatically"

"Elements and equipment which transfer forces or movements to others where they indicate or register the behaviour of an operating function or to actuate another component"

"Intervention in material and energy flows of a machine not directly by hand"

"Influencing of processes which cannot be initiated directly by human intervention"

Definition of control according to DIN 19 226:

"Control means the process in a system in which one or several input variables influence other output variables as a result of the laws pertaining to the system. Controlling is characterized by the open-loop sequence of actions via the single transfer element or the control chain."

This system is regarded for the moment as a self-contained box. The input variables (designated x_e ... in Fig. 1/1) acting on this system are linked in the box and issued as output variables x_a ..., and these output variables now act on the energy flow or mass flow to be controlled.

Fig. 1/1

In general: $x_a = f(x_e)$

The block diagram shown in Fig. 1/2 represents the control itself together with the system to be controlled.

Fig. 1/2

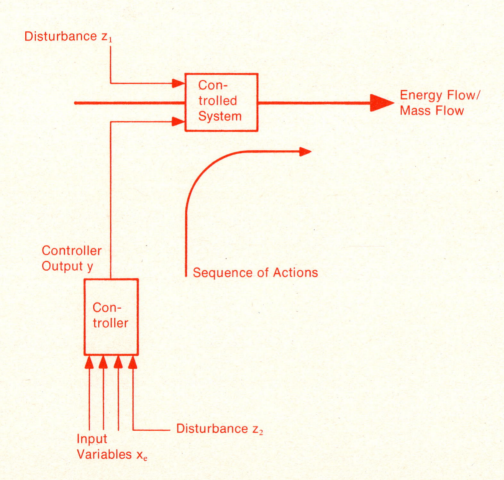

The standard also confirms an extended form of the use of the term "control":

"The word control is often used not only for the process of controlling but also for the complete system in which the control takes place."

In accordance with this definition, Fig. 1/2 can be designated in its entirety by the term "control". The other terms which occur will be explained at another point.

Definition of automatic control according to DIN 19226:

"Automatic control is a process in which one variable to be controlled (controlled variable), is continuously measured and compared with another variable, the command variable, the process being influenced according to the result of this comparison by modifying to match the command variable. The sequence of actions resulting from this takes place in a closed loop, the control loop. The purpose of the closed-loop control is to match the value of the controlled variable to the value specified by the command variable even if perfect equalization is not attained under the prevailing circumstances."

Here, however, the controlled system is influenced by the error resulting from a comparison between the output from the controlled system (i.e. the variable to be controlled) and a specified command variable (the desired value). (Fig. 1/3).

Fig. 1/3

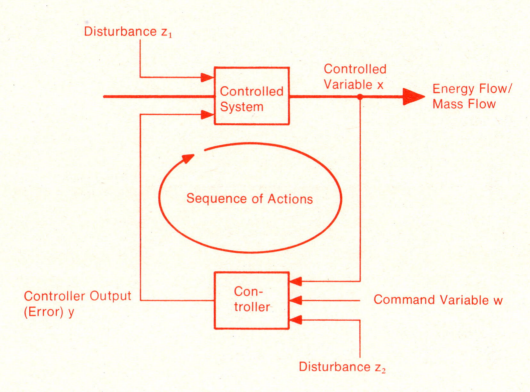

The closed-loop sequence of actions can be recognized clearly here, in contrast to the open-loop sequence of actions in Fig. 1/2.

In the one case one speaks of a closed-loop system and in the other case of an open-loop system. As a result of the arrangement, an important inherent fact can be established: disturbance factors are eliminated in an automatic control system, whereas in an open-loop control they pass through the system without restraint.

1.3 Various terms in automatic control and control engineering

At this point, the more important terms and those occurring most frequently should be listed and explained in greater or lesser detail depending on their difficulty and significance (in accordance with DIN 19226 and DIN 44300 "Informationsverarbeitung, Begriffe" ("Information Processing") April 1965 and also draft August 1968).

1.3.1 Actuating path and direction of action

Actuating path: Path along which the actions determining a control operation are transmitted.

Direction of action: Direction in which the actions are transmitted.

1.3.1.1 Constituent sections of the actuating path

Controlled system: That part of the total system to be influenced.

Actuator: Element which acts on the mass flow or energy flow to be controlled and located at the input to the controlled system.

Controller: Part of the actuating path causing the controlled system to be influenced by the actuator.

The controller is understood to mean the control or automatic control system proper, that is those elements which link the input signals in accordance with the respective laws.

(Not part of the controller: indicating and registering devices, mounted elements, etc.).

Disturbance point: Point at which a factor acts which is not influenced by the system and which disturbs the condition to be maintained.

1.3.1.2 Variables and ranges of variables in the actuating path

Controller output y: Output from the controller and at the same time input variable to the controlled system.

Range of controller output y_h: Range within which the controller output may be adjusted.

Desired value x_A: The value to be acted on by the control.

Control range x_{Ah}: Range within which the desired value may be when the control is operating properly.

Command variable w: Value introduced from outside to the control chain or the control loop and which the output value is to follow in a predetermined manner (e. g. set-point device in closed-loop control, input signal in open-loop control).

Range of command variable w_h: Possible range of the command variable.

Disturbance variable z: Variable acting from outside influencing the intended action of the control.

Disturbance range z_h: Range within which the disturbance variable may be, without adversely affecting the operability of the control.

1.3.2 Signals

Signals represent information. The representation may refer to the value or the change in values of a physical dimension and may refer to transmission, processing or storage of information.

In abstract considerations, the reference to physical dimensions can also be omitted and values and changes in value of mathematical quantities may be called signals. One must distinguish between two types of signals:

Analog signal: Various information is assigned continuously point by point to a range of values.

Digital signal: The range to be considered is divided into a finite number of separate value ranges, and one specific item of information is assigned to each range of values.

The digital signal group includes the binary signal, also known as an on-off signal.

Binary signal: Single parameter signal with only two possible values.

Thus, this is a signal representing two items of information (e. g. ON – OFF, YES – NO, present – not present).

Whereas one operates mainly with analog signals in automatic control, digital signals are used more frequently in control engineering, and the digital signals are mainly in the form of binary signals. These binary signals are of considerable significance for information processing because they can be easily produced by equipment (e. g. switches) and they can also be processed simply. In practice, it is essential to clearly define the relationship between range of values and signal in the case of binary signals, and, in order to avoid any overlapping, to insert a sufficiently large free zone between the two ranges of values, e. g. 0-signal 0 . . . 50 kPa (0 . . . 0.5 bar/0 . . . 7.25 psi), 1 – signal 400 . . . 800 kPa (4 . . . 8 bar/58 . . . 116 psi).

To illustrate the above, the following examples are given.

Example 1:

If a continuously changeable pressure from 0 . . . 600 kPa (0 . . . 6 bar/0 . . . 87 psi) is considered, each intermediate value of the range to be considered may be assigned a specific signal.

If the pressure is indicated on a Bourdon pressure gauge for example, each intermediate value corresponds to a specific position of the pointer. The position of the pointer represents an analog signal. (Fig. 1/4)

Fig. 1/4

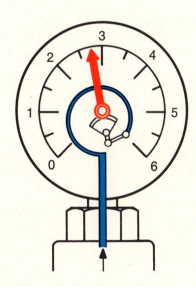

If the dial is now divided into separate value ranges, e.g. in pressure steps of 50 kPa (0.5 bar/7.25 psi), and if each range is assigned a specific item of information, e.g.:

50 ... 100 kPa (0.5 ... 1 bar/7.25 ... 14.5 psi) value 1
100 ... 150 kPa (1 ... 1.5 bar/14.5 ... 21.75 psi) value 1.5
150 ... 200 kPa (1.5 ... 2 bar/21.75 ... 29 psi) value 2

then we are dealing with digital signals.

Example 2:

The flow volume in a tube is influenced through a valve by a continuously changeable pressure p_s. The flow is adjusted by the position of the valve disk. In Fig. 1/5a, p_s is applied to a diaphragm cylinder with spring return.

Within the range of correction, the valve stem may assume any position depending on the control pressure p_s. This is therefore an analog drive. In contrast, the valve stem in Fig. 1/5b can assume only two stable positions. Owing to the upstream poppet valve, which of course opens the through passage at a specific pressure p_s, the single-acting diaphragm cylinder is restrained either in the forward or in the rear end position, and thus displays binary behaviour. This can also be seen clearly on the diagrams.

With a uniformly constant change in the control pressure p_s, the correcting displacement s_2 is changed almost instantly whereas s_1 responds continuously to p_s over the range of correction.

Fig. 1/5a and 1/5b

Fig. 1/5a

Fig. 1/5b

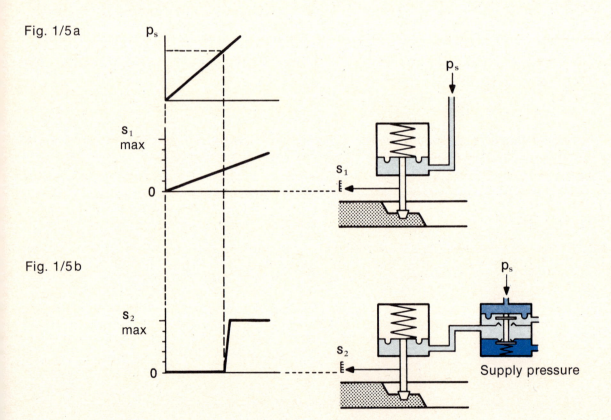

© by FESTO DIDACTIC

1.3.3 Signal flow diagram

By this is meant the symbolic representation of the effective relationships between the signals in a system or in a number of interacting systems.

The elements of a signal flow diagram are:

a) Block and line of action:

Rectangle (block, see Fig. 1/6) for representing the relationship between output signals and input signals.

Line of action (line with arrow) for representing signals as well as their directions.

Fig. 1/6

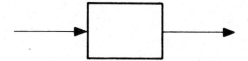

Where known, the effective relationships can be entered in this block (e. g. characteristic, function equation, etc.).

b) Linkage points:

Addition points can be represented by a circle, and the applicable arithmetic sign is written in. (see Fig. 1/7).

Fig. 1/7

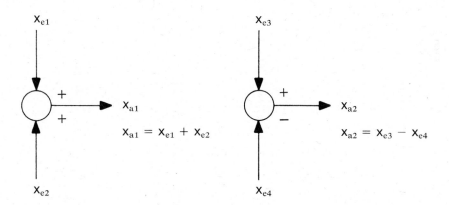

$x_{a1} = x_{e1} + x_{e2}$

$x_{a2} = x_{e3} - x_{e4}$

A change of sign can be represented as shown in Fig. 1/8.

Fig. 1/8

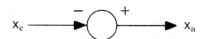

c) Branch point:

At a branch point, the line of action splits into several branches and the signal passes into each branch without any change in magnitude. Fig. 1/9 shows this branching.

Fig. 1/9

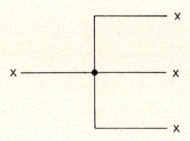

A branch point is represented by a distributor point (diameter approximately treble line thickness).

d) Arrangements:

A signal flow diagram can always be made up from the three basic structures:

- chain structure
- parallel structure
- loop structure

Example 3:

Representation of a closed loop in the signal flow diagram.

Fig. 1/10

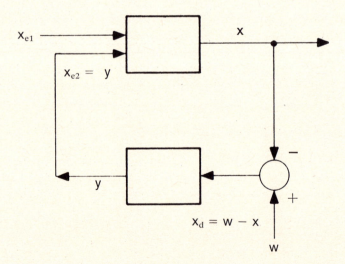

1.3.4 Summary

If the reader wishes to delve deeper into the subject matter offered up to this point, we would recommend him to study the DIN 19 226 and DIN 44 300 Standards.

So far the two fields automatic control and control engineering have been considered as an entity with respect to common terms and their definitions.

The two will now be separated. The following text applies to control engineering and although the subject matter is valid in a general sense it does in part already deal specifically with cylinder controls.

1.4 Breakdown of the control chain

In the preceding sections, the controller has been represented as a self-contained block.

This block can be broken down even further. A control can always be broken down by the same method to show the arrangement of the individual components. At the same time, this shows the signal flow.

Fig. 1/11 shows a breakdown of this kind.

Fig. 1/11

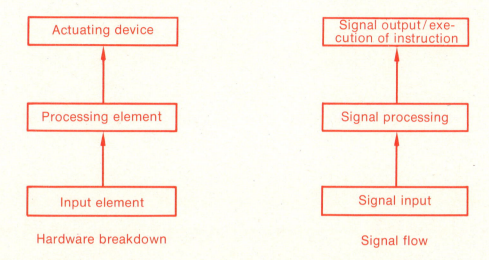

The control chain is thus characterized by a signal flow from signal input via signal processing to signal output/execution of instruction.

In hardware terms, this means that input devices, processing devices and output devices must exist for these signals.

At this point, it is necessary to introduce some hardware terms.

Input element: The first component in the direction of action of a control chain.

Actuating mechanism: This element has a direct effect on a controlled system. The actuating mechanism moves the final control element, provided the latter is mechanically actuated.

Actuating device: Consists of actuating mechanism and final control element.

Signal transformer or transducer: Device which transforms an input signal as clearly as possible into an associated output signal, where necessary using auxiliary energy.
Among others, this group of devices includes amplifiers and signal converters.

Signal amplifier (amplifier): Device with auxiliary energy for power amplification.

Signal converter: Devices in which the input and output signals have a different structure.

Example of such designs are:

Signal elements: Limit switch with cam and roller operation; contacless signallers such as proximity switches, light barriers, air barriers, reflex sensors, etc.; manual push buttons; manual switches; foot switches etc.

Processing elements: Electronic logic elements, contactors, relays, valves reversed by pressure, pneumatic logic elements, etc.

Final control elements: Power contactors, pneumatic and hydraulic valves etc.

Drive elements: Electric motors, pneumatic and hydraulic cylinders, pneumatic and hydraulic motors.

A further distinction between various types of controls can be found when one considers the energy requirements.

Controls without auxiliary energy:

The power required to adjust the final control element is provided by the input element of the control.

Controls with auxiliary energy:

The power required to adjust the final control element is supplied entirely or in part by a source of auxiliary energy.

Furthermore, it is possible to operate with different levels of energy within the control chain whereby one must distinguish between

working energy and control energy.

Working energy is understood to mean that energy which is required to operate the actuating device, and **control energy** is that which supplies signal input and signal processing.

This subdivision does mean however that additional devices must be introduced into the control chain for signal transformation. These devices consist of amplifiers or converters depending on whether or not the same form of energy is used for the operative part and the control part.

Based on these considerations, an extended control chain can now be drawn up. (As shown in Fig. 1/12 with controlled system.)

Fig. 1/12

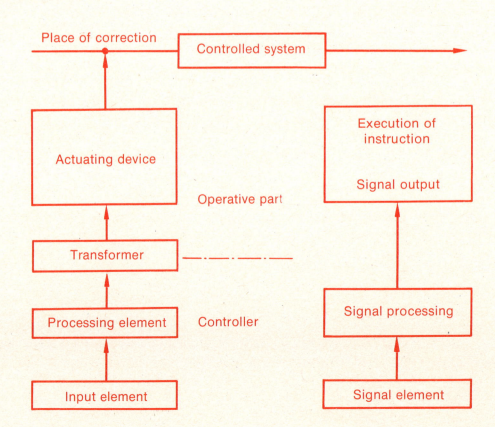

Example 4:

Control with the same type of energy for operative part and control part:

Parts of different width (there being two different widths) approaching on a conveyor belt are sensed by a feeler mechanism and sorted by means of a switching section which is set by means of a pneumatic cylinder (Fig. 1/13). (Distance between parts sufficiently large to prevent overlapping.)

Fig. 1/13

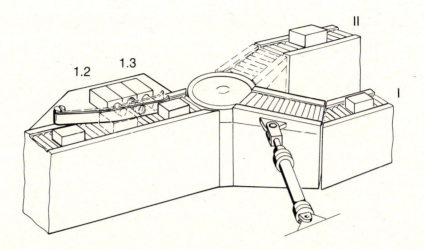

The circuit and the layout of the control chain can be seen in Fig. 1/14.

A thin part causes only 1.3 to be operated, and the parts are taken onto conveyor belt I.

When a wider part is encountered, both valves 1.2 and 1.3 are operated and the parts are taken onto belt II.

Fig. 1/14

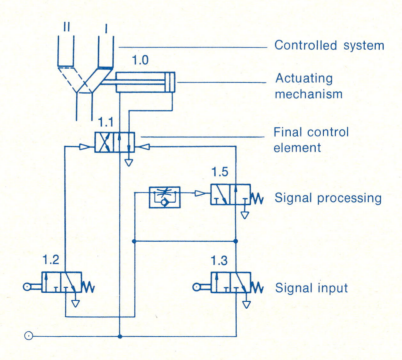

Frequently, when considering and designing a control, only a part of the total system is of interest. It is then possible to take a part out of the overall "control" and to consider this as a partial control.

The hardware used at different stages in the signal flow is shown in Fig. 1/15 as an example for pneumatics and for electrical systems.

Fig. 1/15

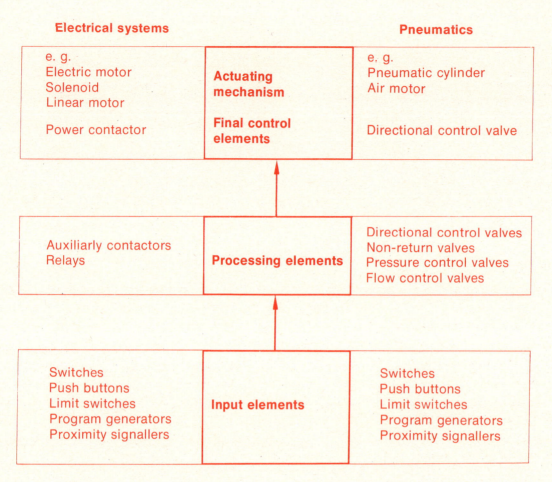

1.5 Types of energy for operative part and control part – summary and definition

Because it is possible by means of suitable devices (signal transformers, transducers) to convert signals of one type of energy into signals of another type of energy, in control engineering one can work within a controlled system with different types of energy.

Thus, it is possible to design a control on the basis of optimum economical and technical aspects.

In practice however, it is not always easy or straightforward to select the "right control system". Apart from the immediate requirements of the problem, the auxiliary requirements in particular (such as for example place of installation, environmental influences, available maintenance personnel etc.) determine the solution. These auxiliary conditions often conflict extremely with the simple solution to the problem and they can make project engineering considerably more difficult.

If a system uses different types of energy for the operative and control parts, one refers to a mixed technology. It should be mentioned at this point that mixed technology is being used to an increasing extent in control design.

The following summary lists the most commonly used working and control media, criteria for selection, and also the advantages and disadvantages.

It is not claimed that all relevant facts are included in the list, nor would it be possible to do this. Only the most important points are listed in summarized form.

1.5.1 Working media:

- Mechanical (use limited, will not be dealt with in any detail here)
- Eletrical
- Hydraulics
- Pneumatics

Criteria for system selection

- Force
- Displacement
- Type of motion (linear, rotary etc.)
- Speed
- Physical size
- Life
- Sensitivity
- Working safety

Characteristics:

Electrical:

Energy storage difficult, energy transmission good and fast (almost speed of light), energy costs low.

Creation of straight-line motion:

Complex and expensive, as it is necessary either to convert by mechanical means or short displacements possible with lifting magnets and only small forces possible with comparable linear motors. Large physical size.

Creation of rotary motion:

Very high efficiency, large physical size, speed limited, favourable characteristic, speed and torque regulation difficult and elaborate.

General:

Elements not overload-proof, or only by elaborate means.

Not intrinsically explosion-proof.

Hydraulics:

Storage of energy only to a limited degree, limited and slow energy transmission (distance which can be covered $\sim 10^2$ m, speed $\sim$ 2–6 m/s), high energy costs.

Creation of straight-line motion:

Very simple, working speed not too high (up to 0.5 m/s max.), very small dimensions, large to very large forces can be achieved.

Creation of rotary motion:

Simple, not very high speeds can be obtained, speeds constant even in the low range, high efficiency, high torques.

General:

Elements are overload-proof, line installation difficult and expensive as one is working with high pressure and it must be ensured that the system is completely sealed.

Pneumatics:

Storage of energy presents no problems, limited and slow energy transmission (distance which can be covered $\sim 10^3$ m, speed $\sim$ 20–40 m/s), very high energy costs.

Creation of straight-line motion:

Simple and cheap, high working speeds (normally 1–2 m/s), stroke length limited, up to 2 m depending on the design, force obtainable is limited (up to $\sim$ 40 000 N max., normally up to 10 000 N), small dimensions.

Creation of rotary motion:

Simple and cheap, high operating costs due to poor efficiency, high speeds (up to 500 000 min^{-1}). Torque obtainable not too high.

General:

Elements overload-proof, intrinsically explosion-proof, very simple regulation of speed, torque, working speed, and feed force.

1.5.2 Control media

- Mechanical
- Electrical (electromechanical)
- Electronics
- Normal-pressure pneumatics
- Low-pressure pneumatics
- Hydraulics

Criteria for system selection

- Signal speed
- Switching times of elements
- Working safety of elements
- Life
- Sensitivity to environmental influences
- Space requirement
- Ease of maintenance

	Electrical	Electronics	Normal-pressure pneumatics	Low-pressure pneumatics
Signal speed	very high ~ speed of light		approx. 40–70 m/s	100–200 m/s normal, to some extent speed of sound
Distance which can be covered	practically unlimited		limited by speed of signal	
Switching times of elements	> 10 ms	≪ 1 ms	> 10 ms	> 10 ms
Reliability	sensitive to environmental influences such as dust, humidity etc.	very sensitive to environmental influences such as dust, humidity, interference fields	very insensitive to environmental influences, with clean working air very long life	insensitive to environmental influences, sensitive to contaminated air
Space requirement	large	very small	very large	small
Main type of Signal processing	digital	digital, analog	digital	digital, analog
Components	contactors, relays	electronic valves, transistors	directional control valves	static, dynamic elements

1.6 Differentiating characteristics of controls

Controls can be classified according to various criteria:
e. g.

1.6.1 According to the control energy used

The important points have been dealt with in chapters 1.4 and 1.5.

1.6.2 According to the mode of operation with respect to signal processing

Two terms are common here:

Combined controls:

Signal processing is of the positive kind. A certain combination of input signals is always associated with a specific combination of output signals. These controls operate without time behaviour.

Sequential controls:

These include all controls containing elements with time behaviour (e. g. timing elements, storage devices, etc.).

This classification has been included here because these terms are still encountered frequently in publications. The types of control described here can however be easily accommodated in the standard classification (shown in 1.6.3).

1.6.3 According to the type of operating sequence

Types of control according to DIN 19226:

Fig. 1/16

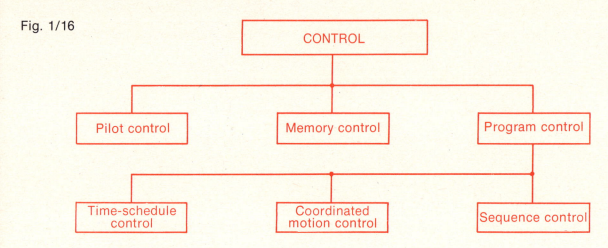

A control must be identified with one of the main groups (Fig. 1/16) and the allocation depends on the particular problem. If a program control is involved, the project engineer may select from the three subdivisions of program control, i.e. a program control can be built up in different ways.

1.6.4 Pilot control

There is always a unique relationship between command variable and output variable.

Example 5:

Copying on machine tools. The movements of the tracer pin are uniquely related to the movements of the tool.

Example 6:

Brightness control. The brightness of the lamp is at all times uniquely related to the position of the resistor or of the transformer in Fig. 1/17.

Fig. 1/17

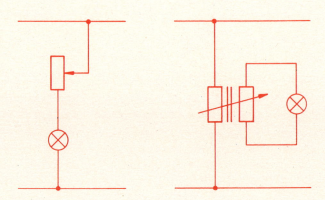

1.6.5 Memory control

After removing or taking back the command variable, the value reached by the output variable is retained until an opposing signal is presented.

Example 7:

Switch-on and switch-off of an electric motor by means of a pressure switch.

Fig. 1/18

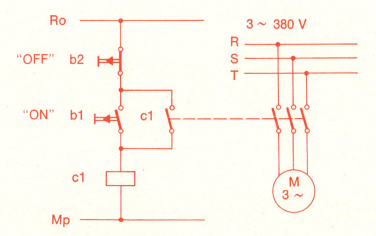

If the ON button (b1) in Fig. 1/18 is operated, contactor c1 pulls up and, owing to the holding circuit, continues to be connected electrically even when b1 is released. The obtained condition – motor running – is thus preserved until an opposing signal, in this case provided by operating the OFF button b2, is input into the control, and then the new condition – motor standstill – is again maintained until the opposing signal is provided.

Example 8:

Controlling the advance and return movement of a pneumatic cylinder by means of a manual switch (Fig. 1/19).

In this case also, the condition obtained by the output variable (in this case the position of the cylinder) is retained after the command variable has been removed (operation of 1.2 or 1.3) until the opposing signal is presented.

Fig. 1/19

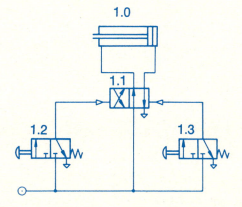

1.6.6 Program controls

The following types of control fall into this category.

25

1.6.6.1 Time-schedule control

The command variables are supplied by a time-dependent program transmitter (program storage device).

A time-schedule control is thus identified by the presence of a program transmitter and a time-dependent operating sequence of the program.

A program transmitter may be a:

- Camshaft
- Cam disk
- Program belt
- Punch card
- Punch tape
- etc.

Example 9:

A player piano displays all the characteristics of a time-schedule control. The program is contained in the program transmitter, which in this case is either a punch tape or a drum, and is run through on a time-dependent basis. (Constant speed of the program transmitter drive motor.)

1.6.6.2 Coordinated motion control

The command variables are provided by a program transmitter, the output variables of which are dependent on the distance covered. (Displacement or position of a movable part of the controlled system.)

Example 10:

Movement of a double-acting cylinder. The advance motion is tripped by operating the START button 1.2, the return motion being effected by a limit switch after a certain length of travel (Fig. 1/20). In this case, program input depends on the position of the limit switch 1.3.

Fig. 1/20

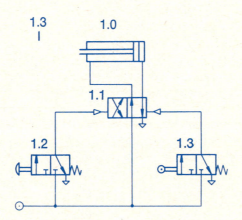

© by FESTO DIDACTIC

1.6.6.3 Sequence control

The operating sequence program is stored in a program transmitter which runs through the program step-by-step in accordance with the condition reached at any one time by the controlled system. This program may be fixed, or it may be variable and specified by means of punch cards, punch tapes, magnetic tapes, or other suitable storage media.

A sequence control is identified by having a program transmitter and also equipment which is capable of interrogating the conditions prevailing in the system.

Example 11:

Punch tape control of a machine tool. An instruction e. g. tool slide forward, is provided via the punch tape. The "status" of the machine is interrogated by means of a limit switch in the end position of the tool slide. If the tool slide has reached this position, the limit switch signal causes the punch tape to be moved forward by one step to allow input of the next instruction, and the tape remains stationary at this position until a new status message is received.

1.6.7 Comparison and definitions of program control systems

The type of control used in a particular requirement for a program control depends solely on the problem at hand, the requirements imposed, environmental influences and the auxiliary conditions. In this case also, no generally valid recommendations can be made for the application, and at best one can only define particular characteristics for each system. The following represents a summary:

Time-schedule control:

Centrally stored program; simple to program; usually, compact construction; simple connection; time-constant program execution; program execution insensitive to disturbing factors and independent of machine operating sequence; no check on the particular working sequence, hence no operating sequence reliability. Disturbances in the machine operating sequence have no effect on the program execution.

Coordinated motion control:

Program defined by the arrangement of limit switches and signal elements, hence the layout is not clearly arranged and is not easy to service, difficult to install.
Operating sequence reliability provided by displacement-dependent sequence, disturbances in the machine operating sequence are registered, program execution may be interrupted.

Sequence control:

The advantages of a central program transmitter (see time-schedule control) and the advantage of a check of the machine statuses (operating sequence reliability) are combined. However, this type of control requires that both program transmitter and equipment for checking the momentary status exist and furthermore that it is possible to advance the program transmitter step-by-step (e. g. stepping motor).

1.7 Means of representing motion sequences and switching conditions

It goes without saying that movement sequences and switching conditions of working and control elements must be represented in a clear fashion. As soon as one is confronted with a rather more difficult problem, the relationships can be identified quickly and with certainty only if a suitable form of representation has been selected. Furthermore, only neat representation allows large-scale projects to be understood clearly.
An example will be used to show the most common forms of representation. At this stage, the details of the application will not be mentioned, or at least only very briefly.

Example 12:

Packages arriving on a roller-conveyor (Fig. 1/21) are lifted by a pneumatic cylinder A and pushed onto another conveyor by a second cylinder B. The problem requires that cylinder B return only when A has reached the rear end position.

Fig. 1/21

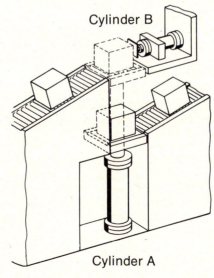

Cylinder B

Cylinder A

Means of representing the working sequence for example 12:

1.7.1 Writing down in chronological sequence

Cylinder A moves out and lifts the packages,
Cylinder B pushes the packages onto conveyor II,
Cylinder A travels down,
Cylinder B travels back.

1.7.2 Tabular form

Working step	Motion cyl. A	Motion cyl. B
1	advance	–
2	–	advance
3	return	–
4	–	return

1.7.3 Vector diagram

Simplified representation
Advance motion represented by →
Return motion represented by ←

A →
B →
A ←
B ←

1.7.4 Abbreviated notation

Designation for advance motion: +
Designation for return motion: –

A +, B +, A –, B –   or   A+
                                             B+
                                             A–
                                             B–

1.7.5 Graphical representation in diagram form

First, a few introductory words:

The VDI Regulation 3260 (February 1963 edition) is concerned among other things with the means of representing functional sequences of working machinery and production plant. However, this regulation is presently being completely revised, and among other things it also still makes use of the old definition of terms contained in the DIN Standard 19 226.

For these reasons, the possible means of representation given are put together and built up in accordance with the requirements of circuit design for pneumatic and electro-pneumatic controls, and not always exactly in accordance with the current standards.

In representing functional sequences, one must differentiate between

– Motion diagram
– Control diagram. } Function diagram

Whereas the motion diagram records conditions relating to working elements and components, the control diagram provides information concerning the condition of individual control elements. When both diagrams are considered together, they are referred to as a function diagram.

1.7.5.1 Motion diagrams

Displacement-step diagram

The operating sequence of a working element is represented by this diagram. The displacement is recorded in relation to the various steps (step: change in condition of any component). If a control has several working elements, these are represented in the same manner and drawn one beneath the other. The relationship is provided by the steps.

Fig. 1/22

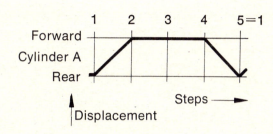

For a pneumatic cylinder A for example, the displacement-step diagram shown in Fig. 1/22 applies:

From step 1 to step 2, the cylinder travels out from the rear to the forward end position, the latter having been reached at step 2. From step 4, the cylinder again returns and has reached the rear end position at step 5.

The displacement-step diagram for example 12 shown in Fig. 1/21 is given in Fig. 1/23.

Fig. 1/23

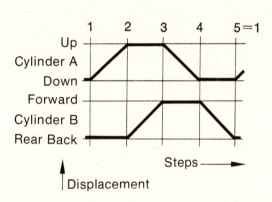

Recommendations for layout when drawing:

– The steps should where possible be drawn linearly and horizontally.
– If possible, the displacement should not be drawn to scale, but of equal size for all components.
– With several units, the distance between the displacements should not be made too small (approx. 1/2 to one step).
– If the condition of the system changes during the motion, e. g. by operating a limit switch while the cylinder is in mid-position or due to change in the advancing speed, intermediate steps may be introduced.
– The steps may be numbered as required.
– The designation of the condition is optional. This may be as in the example above by specifying the position of the cylinder (rear – forward, up – down etc.) or by making use of binary digits (e. g. 0 for rear end position, L or 1 for the forward end position).
– The designation of the unit concerned must be written on the left of the diagram.

Displacement-time diagram

The displacement of a component is drawn in relation to time. In contrast to the displacement-step diagram, in this diagram the time is drawn linearly and establishes the relationship between the individual components.

Displacement-time diagram for example 12:

Fig. 1/24

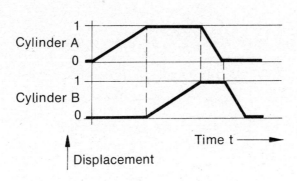

The rules for drawing the diagram are roughly the same as for the displacement-step diagram. The relationships to the displacement-step diagram is indicated by the broken lines (step lines), although the distance between them is now no longer to scale.

Whereas the displacement-step diagram allows the relationships to be seen more clearly, overlaps and varying working speeds can be shown better in the displacement-time diagram.

If diagrams are to be made for rotating working elements (e. g. electric motors, air motors), the same basic procedure should be followed. However, the changes in condition with respect to time are no longer accounted for, i. e. in the displacement-step diagram a change of condition (e. g. switching on an electric motor) does not extend over a whole step but is entered directly on the step line.

1.7.5.2 Control diagram

In the control diagram, the switching condition of a control element is shown in relation to the steps or the times, the switching time itself not being considered, e. g. condition of a relay b1 being opened (Fig. 1/25).

Fig. 1/25

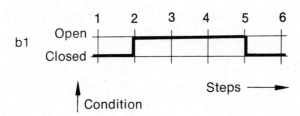

The relay pulls up at step 2 and drops off again at step 5.

Another method of drawing the control diagram is by drawing only on the line (Fig. 1/26).

Fig 1/26

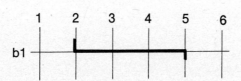

Transfer to the motion diagram is by means of the steps or by means of the time.

The following is recommended:

- The control diagram should if possible be drawn in conjunction with the motion diagram.
- Steps or times should be entered linearly and horizontally.
- Height and separation are optional, but should be selected to ensure clarity.

The combined motion and control diagram for example 12 is shown in Fig. 1/27. The control diagram shows the conditions of the directional control valves which control the cylinder (1.1 for A, 2.1 for B), and the condition of a limit switch 2.2 which is installed at the front end position of cylinder A.

Fig. 1/27

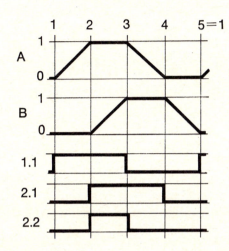

As already mentioned, the switching times of the components are not considered in the control diagram. However, as shown in Fig. 1/27 against limit switch 2.2, the actuating lines for limit switches should be drawn in before or after the step line because in practice the actuating does not occur exactly in the end position but either just before or just after this position. This method of drawing makes the situation perfectly clear as it can be seen beyond doubt that the signal already exists at the step in question, i. e. it already exists completely at the step line.

1.7.6 Symbols and representation standards

Fig 1/28 shows some important symbols and terms based on VDI Regulation 3260 and DIN 55 003.

These symbols may be applied both to drawings and diagrams as well as labels without text on machine tools (see DIN 55 003).

© by FESTO DIDACTIC

Fig. 1/28

Motions

Straight-line motion in direction of arrow

Straight-line motion in two directions

Straight-line motion in direction of arrow, limited

Straight-line motion in direction of arrow, limited, single reciprocation

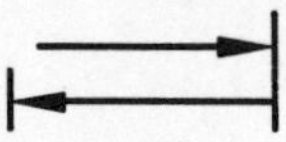

Straight-line motion in direction of arrow, limited, continuous reciprocation

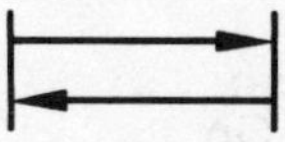

Rotary motion in direction of arrow

Rotary motion in two directions

Rotary motion in direction of arrow, limited

Revolutions / continuous operation / continuous cycle

One revolution / single sequence / single cycle

Revolutions / minute

General symbols

Pressure gauge to DIN 2481

Electrical instrument to DIN 40 716

Electric motor

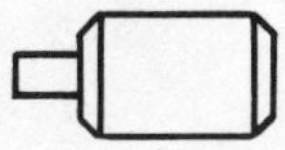

Command symbols

Positioning, establishment of working position

Clamping, chucking

Unclamping

Locking

Unlocking

On

Off

On/off

Push button switching (switched on as long as the button is pressed)

All systems off (emergency switch), colour red

Types of energy

Hydraulic

Pneumatic

Mechanical

Electrical

© by FESTO DIDACTIC

Colour coding of push buttons and lamps (to DIN 43 605)

General:

Colour RED: Elimination of sources of danger
Colour GREEN: Not for indicating a switched-on state

Code:

Colour	Push buttons	Indicators
RED	STOP EMERGENCY STOP	Switched-on state
YELLOW	Start for first cycle	Disturbance
BLACK	Start single movement	
GREEN		Switched-off state
BLUE		Acknowledgment

© by FESTO DIDACTIC

Colour coding of push buttons and lamps (to DIN 43 605)

General:

Colour RED — Illumination of sources of danger
Colour GREEN — Not for indicating a switched-on state

Code:

Colour	Push buttons	Indicators
RED	STOP EMERGENCY STOP	switched-on state
YELLOW	Start for first cycle	Disturbance
BLACK	Start single movement	
GREEN		Switched-off state
BLUE		Acknowledgment

© by FESTO DIDACTIC

1.8 Working out a control problem

The sequence given here for working out a control problem has been found in many cases to be practical. The points listed must be clarified for each problem, and the various facts must be established and noted.

1.8.1 Problem definition, determination of conditions

Right at the start, the problem, and the objectives especially, must be defined clearly and unambiguously. Also, it is very important to list the auxiliary conditions with respect to, for example,

— Ease of operation
— External safety of the system
— Reliability of performance
— Etc.

To ensure that the expressions used are uniform, the following terms and the classification associated with this should be defined:

Auxiliary conditions:

— Auxiliary conditions for functional sequence:
   a) Start conditions
   b) Setting-up conditions
   c) Safety conditions
— Auxiliary conditions for operating influences:
   a) Environmental influences, place of installation
   b) Supply
   c) Personnel

Possible auxiliary conditions for the functional sequence:

Start and setting-up conditions:

AUTOMATIC operation: AUT

Single cycle	↻	One sequence of operations
Continuous cycling	○	Continuous operation
Jogging operation		Step-by-step cycling following the sequence of the sequence of motions

MANUAL operation: MAN

Setting up: Each working element can be operated separately in any sequence
Setting: By operating the "set" button, the system is brought into a defined position.

Safety conditions:

EMERGENCY STOP: The position of the working elements assumed when the EMERGENCY STOP condition applies must be unambiguously defined beforehand.

EMERGENCY STOP unlocking: The system is again released for further operation.

1.8.2 Working energy, working elements

The working energy should be the next topic to be considered. In particular, the operating conditions, as already established in chapter 1.5, should be considered here.

If the working energy has been established, the working elements can be selected and dimensioned. The criteria for selection and calculations applicable to the particular type of equipment used must be applied here.

The working energy used is of course of prime importance to the project engineer designing the control. In many cases, the dimensioning is only of secondary or global interest (e. g. in the case of a pneumatic cylinder whether the stroke and force allow a limit switch to be operated).

1.8.3 Positional sketch

It is always advisable to make a positional sketch based on the description of the problem. This need only be a rough sketch. It should make it easier to identify the factors involved with the working elements, their arrangement, and possibly also their method of working. This positional sketch can of course also be used as a basis for discussion.

1.8.4 Determination of the sequence of operations

At this point, the sequence of functions should be listed. The forms of representation dealt with in chapter 1.7 should be used here. The form chosen depends on the particular problem. It may be of value to make use of several types of representation alongside each other (e. g. motion diagram and tabular list for draft design and calculation).

1.8.5 Choice of type of control

The criteria for selection are given in chapter 1.6.

1.8.6 Choice of control energy

This topic has also been discussed (see chapter 1.5). It is emphasized again at this point that the operating conditions are extremely important (environmental influences, availability of maintenance personnel, etc.).

1.8.7 Cicuit diagram

When all the above points have been determined and clarified, one may start to draw up the circuit diagram. The sequence to be adopted now depends very much on the control energy used. This calls for accurate knowledge of the technology of the equipment, the symbols provided for this, and the behaviour and interaction of the various components in the particular technology selected if one is to end up with a good and reliable cicuit.

1.9 Exercise

All the facts which are required to draw up a circuit diagram are to be put together in accordance with the considerations discussed up to this point for example 13 which is given below.

(It is advisable to proceed as described in chapter 1.8). Missing data may be selected freely.

Example 13:

A fixture, or special machine, is to be designed such that rectangular parts (Fig. 1/29) measuring 80 x 60 x 50 can be stamped on one side, the operating sequence being fully automatic.

Fig. 1/29

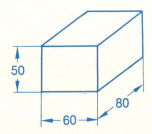

Material:	Aluminium alloy
Required stamping force:	Approx. 800 N
Quantity:	Approx. 8000 parts/day
Weight of punch:	Approx. 80 N

1.9.1 Procedure for working out the exercise

Of course, different solutions can be offered for this exercise. One of the possibilities is given here.

1.9.1.1 Definition of problem and conditions

Working operations to be formed:

- Stack parts (gravity feed magazine)
- Feed in parts (push)
- Hold parts (clamp)
- Work parts (punch)
- Eject parts

Determination of the auxiliary requirements: e. g.

a) Start the system by means of manual "START" button.

b) Selector switch "single cycle", "continuous".
   Position "single cycle": One working cycle is to be performed, then stop in starting position. Position "continuous": After operating the "START" button, fully automatic operation until opposing signal "single cycle" given.

c) Magazine sensing: If the magazine is empty, the system should stop at the starting positon, and it should not be possible to start again until the magazine has been filled.

d) EMERGENCY STOP: When operating the "EMERGENCY STOP" button, the system must be go back immediately to the starting position and be ready to start again only after the "EMERGENCY STOP" button has been unlatched.

1.9.1.2 Selection of working energy and dimensioning of working elements

The operations to be performed can all be accomplished with straight-line movements.

Forces required:	small (punching force 800 N max.)
Length of movement:	max. 200–300 mm
Working speed:	with 8 hours operating time, approx. 3,6 sec./part
Working energy selected:	pneumatic

Working elements required:

Feed-in cylinder   A
Clamp cylinder   A
Punch cylinder   B
Eject cylinder   C

If the arrangement is suitable, the feeding and clamping operations can be performed by one cylinder.

Dimensioning of the working elements:

It is assumed that it is known how to solve this part of the problem.
Forces and strokes are sufficient to operate limit switches. With a view to working safety and working speed, all three cylinders are made double-acting.

1.9.1.3 Positional sketch

Fig. 1/30

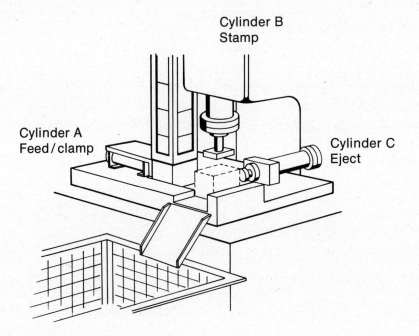

1.9.1.4 Determination of sequence of operations

Working sequence:

Push in   A
Clamp   A
Stamp   B
Unclamp A
Eject   C

Sequence of motion: e. g. abbreviated notation

A+
B+
B−
A−
C+
C−

Displacement-step diagram:

Fig. 1/31

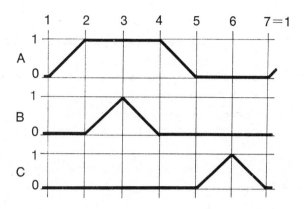

1.9.1.5 Selection of type of control

Identification of main group:

This is a program control (control with an automatically performed working program in accordance with specific rules).

Selection of type of program control: For the example given here, coordinated motion control

- Certainty of operation.
- For the scope of the problem, probably the cheapest solution (no program transmitter).
- No change of program necessary.

1.9.1.6 Control energy

With regard to working media and scope of the problem, there are two possibilities:

pneumatics and electrical,

and in this case a completely pneumatic solution is probably the most favourable.

(Only one form of energy for working and controlling, hence no converter necessary, only one supply of energy to the machine, high degree of operating reliability, insensitive, etc.)

The operating conditions will however also be decisive here: e. g.

- which maintenance personnel are available
- environment in which the machine is located

Selected: pneumatics.

Part 2: Pneumatic Controls

2.1 Introduction

In its conventional sense, the term "pneumatics" is today no longer adequate to describe and define clearly the entire field of working and controlling with air. The fields of application have become too diverse. At the same time, the advent of new techniques has split up the common area of application. For these reasons, it is necessary here to briefly outline the boundaries.

Although many designations exist for the various sectors of pneumatics (pneumatics in general meaning the industrial application of air as a working and controlling medium), there is as yet no prospect of a common terminology.

Some definitions are therefore given below, arbitrary to a greater or lesser degree, with a view to clarifying some of the terminological confusion.

One criterion for grouping is a breakdown into pressure ranges.

Low-pressure pneumatics:

(also known as fluidics, fluid logic, etc.)
Pressure range: ~ up to 150 kPa (1.5 bar/21.75 psi)

This branch covers all systems which are available solely for control purposes in the specified pressure range.
As far as equipment is concerned, it is immaterial whether the elements are static, semi-static, or dynamic.

Normal-pressure pneumatics (conventional pneumatics):

Pressure range: 150 . . . 1600 kPa (1.5 . . . 16 bar/21.75 . . . 232 psi)

Covers the branch of conventional pneumatics, i. e. working and control elements in the specified pressure range.

High-pressure pneumatics:

Pressure range: over 1600 kPa (16 bar/232 psi)

This branch covers mainly special usages of pneumatics in the working part.

With one exception, the following deals with controls in the field of normal-pressure pneumatics. Since non-contacting pneumatic switching elements – nozzles, air gates – are acquiring increasing significance in control engineering, it will be shown here how these low-pressure pneumatic elements can also be used for and incorporated in normal-pressure pneumatic controls.

2.2 List of symbols used

A list of the symbols (assumed to be known) used in pneumatics is given on the following pages (based on DIN 24 300, draft November 1973 "Oil Hydraulics and Pneumatics, Designations and Symbols").

Only these symbols will be used in this fundamental control engineering course, or alternatively symbols whose origin is based on this standard and which have been built up on the same basic principles.

Pneumatic symbols

DIN 24 300
Draft November 1973
and non-standardized special symbols

Energy conversion

Compressor

Vacuum pump

Pneumatic constant motor with one direction of flow

Pneumatic constant motor with two directions of flow

Pneumatic motor with adjustable displacement volume, 1 direction of flow

Pneumatic motor with adjustable displacement volume, 2 directions of flow

Pneumatic motor with limited range of swivel

Single-acting cylinder return stroke by external force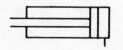

Single-acting cylinder
return stroke by spring

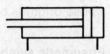

Double-acting cylinder with
single-ended piston rod

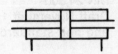

Double-acting cylinder with
double-ended piston rod

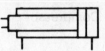

Differential cylinder with
single-ended piston rod

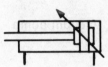

Double-acting cylinder with
cushioning adjustable at both
ends

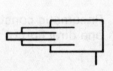

Single-acting telescopic cylinder,
return stroke by external force

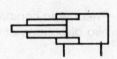

Double-acting telescopic
cylinder

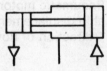

Pressure intensifier for the same
fluid

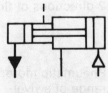

Pressure intensifier for air
and liquid

Pneumatic-hydraulic actuator
e. g. from air – liquid

44

**Energy control and regulation
Directional control valves**

2/2-way valve
closed normal position

2/2-way valve
open normal position

3/2-way valve
closed normal position

3/2-way valve
open normal position

3/3-way valve
closed neutral position

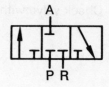

4/2-way valve

4/3-way valve
closed neutral position

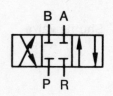

4/3-way valve
floating neutral position

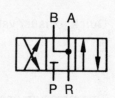

5/2-way valve

5/3-way valve
closed neutral position

Directional control valve with
intermediate switching positions
and two final positions

Directional control valve; simplified
representation with (as an example)
4 ports

Non-return valves

Check valve without spring

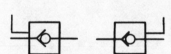

Check valve with spring

Pilot-controlled check valve

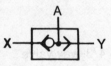

Shuttle valve

Quick exhaust valve

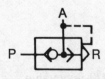

Two-pressure valve
(not standardized)

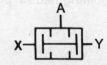

© by FESTO DIDACTIC

Pressure control valves

Pressure relief valve, adjustable

Sequence valve, adjustable

Sequence valve with pressure relief
(3-way function), adjustable
(not standardized)

Pressure regulator, adjustable

Relieving pressure regulator,
adjustable

Flow control valves

Throttle valve with constant
restriction

Diaphragm valve with constant
restriction

Throttle valve, adjustable, any
type of operation

Throttle valve, adjustable,
manual operation

Throttle valve, adjustable,
mechanical operation against
return spring

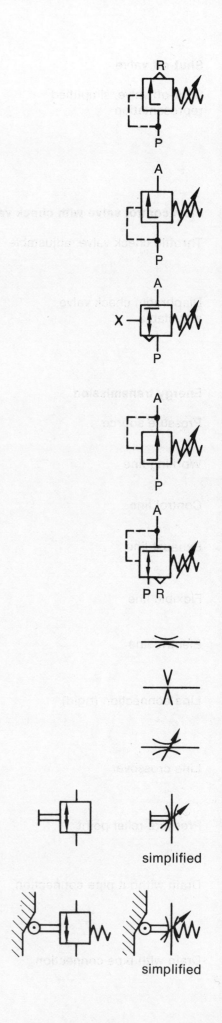

simplified

simplified

Shut-off valve

Shut-off valve, simplified representation

Flow control valve with check valve connected in parallel

Throttle check valve, adjustable

Diaphragm check valve, adjustable

Energy transmission

Pressure source

Working line

Control line

Exhaust line

Flexible line

Electric line

Line connection (rigid)

Line crossover

Pressure relief point

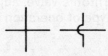

Drain without pipe connection

Drain with pipe connection

Air connection point, closed

Air connection point with connected line

Quick release coupling, without mechanically opened non-return valves, coupled

Quick release coupling, with mechanically opened non-return valves, coupled

Quick release coupling, uncoupled, line open

Quick release coupling, uncoupled, line closed by non-return valve

Rotary connection with 1 path

Rotary connection with 2 paths

Silencer

Pneumatic capacitor

Filter

Water trap, manually operated

Water trap with automatic drain

Filter with water trap, automatic

Dryer

Lubricator

Air service unit (filter, pressure regulating valve, lubricator, and pressure gauge), simplified representation

Cooler

Control mechanisms
Mechanical components

Shaft, rotary motion in 1 direction

Shaft, rotary motion in 2 directions

Detent

Locking device (* symbol for the control method for releasing the locking device)

Over-centre device

Hinge joint, simple

Hinge joint with extended lever

Hinge with fixed pivot

© by FESTO DIDACTIC

**Control methods
Manual controls**

General

Button

Lever

Pedal

Mechanical controls

Plunger

Spring

Roller plunger

Roller plunger with idle return

Sensing element (not standardized)

Electrical controls

Solenoid with 1 effective winding

Solenoid with 2 windings acting in opposition

Electric motor with continuous rotary motion

Electric stepping motor

Pressure controls

Direct by application of pressure

Direct by pressure relief

Differential pressure actuation

Pressure centred

Spring centred

Indirect by application of pressure (piloted)

Indirect by pressure relief

By application of pressure through intensifier (not standardized)

By application of pressure through amplifier and indirectly (not standardized)

By application of pressure, type of control produces alternating behaviour (not standardized)

Combined controls

Solenoid and pilot valve

Solenoid or pilot valve

Solenoid or manual operation with return spring

General;
*: explanatory symbol (specify in footnote)

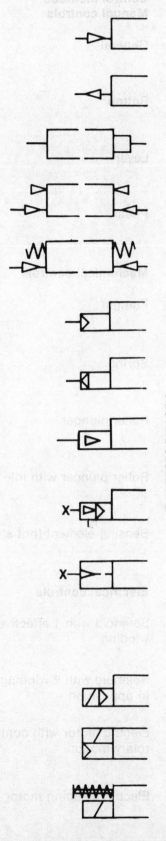

© by FESTO DIDACTIC

Other equipment

Pressure measuring instrument

Differential pressure instrument

Temperature measuring instrument

Flow measuring instrument (flow)

Flow measuring instrument (volume)

Pressure switch

Pressure probe

Temperature probe

Flow probe

Indicator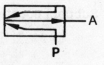

**Special symbols
proximity circuit elements**
(not standardized)

Reflex sensor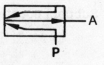

Nozzle, general, emitter nozzle for air gate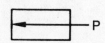

Collector nozzle with air supply,
for air gate

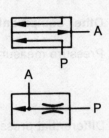

Back-pressure nozzle

Interruptible jet sensor

Amplifiers

Amplifier (e. g. from 50 Pa to 10 kPa)

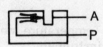

Flow amplifier

3/2-way valve with amplifier
(e. g. from 10 kPa to 600 kPa)

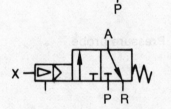

© by FESTO DIDACTIC

Signal converters
(not standardized)

Electrical-pneumatic

Pneumatic-electrical

Counters (not standardized)

Subtract counter

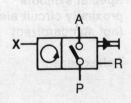

54

Add-and-subtract counter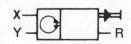

Add counter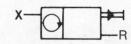

Designation of connections

A, B, C ...	Working lines
P	Supply air, compressed air connection
R, S, T ...	Drain, exhaust points
L	Leakage line
Z, Y, X ...	Control lines

To CETOP RP68 (draft)

2, 4, 6 ...	Working lines
1	Supply air, compressed air connection
3, 5, 7 ...	Drain, exhaust points
9	Leakage line
12, 14, 16, 18 ...	Control lines

2.3 Diagrammatic representation of pneumatic circuit diagrams

Recommendations are contained in the VDI Standard 3226 – Pneumatic Circuits, Circuit Diagrams – (December 1966 issue presently being revised). The types of representation used in the following text are similar to this standard but are not identical in all respects.

2.3.1 Building-up the circuit diagram

The layout should correspond to the control chain flowchart, i. e. there should be a signal flow from the bottom to the top (Part 1, Chap. 4). As the energy supply is of course significant for the circuit diagram, it must also be included in the flowchart – all elements required for the energy supply should be drawn in at the bottom, and the energy should be distributed from the bottom to the top.

In larger circuits, the entire energy supply section (service unit, shut-off valve, various distribution connections etc.) can be drawn separately to one side of the drawing in order to simplify it, and the energy connections to the various elements can be shown by means of ⊙ . Fig. 2/1 shows this breakdown.

Fig. 2/1

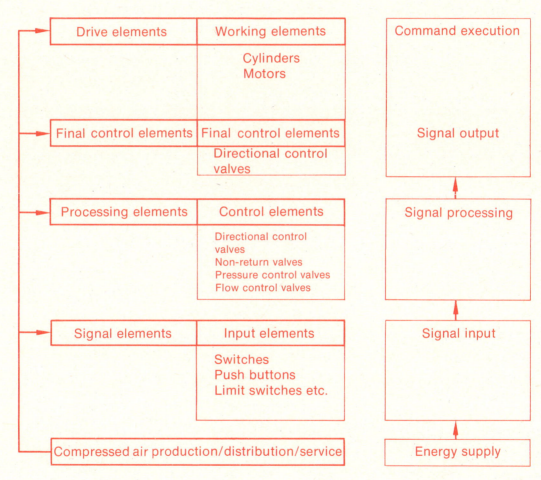

This layout means that the circuit diagram must be drawn without considering the actual physical locations of the elements and it is recommended that all cylinders and directional control valves be drawn horizontally.

Example 14:

It is required that the piston rod of a double-acting pneumatic cylinder travel out if either a manual button or a pedal is operated and that it return to its starting position after reaching the forward end position. (When the signal element which initiates the outward travel is no longer operated.)

Fig. 2/2

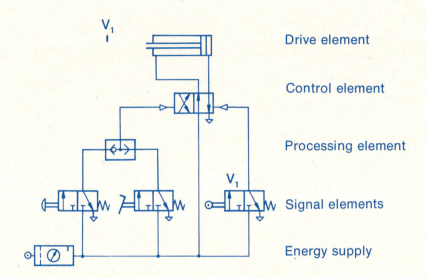

Drive element

Control element

Processing element

Signal elements

Energy supply

Apart from the layout in accordance with the control chain flowchart, the circuit diagram in Fig. 2/2 illustrates the difference between the drawn positions and the actual physical locations. In practice, valve V_1 is mounted at the front end position of the cylinder.

As this part is a signal element, however, it is drawn at the bottom in the circuit diagram. To identify the actual physical arrangement, the true position is indicated by a broken line; where the signal is given in one direction only, e. g. valve with idle return roller, this is shown by an arrow head pointing in the direction of the signal flow.

If the control is complex and contains several working elements, the control should be broken down into separate control chains in which a chain can be formed for each working element.
Wherever possible, these chains should be drawn next to each other in the same order as the movement sequence.

2.3.2 Designating the elements

Two types of designation have been found to be suitable and are frequently encountered:

– designation using digits
– designation using letters

2.3.2.1 Designation using digits

Here too, there are many possibilities for designating with digits. 2 systems are used frequently:

a) serial numbering
This system is advisable for complex controls and particularly in such cases where possibility b) cannot be considered because of overlapping.

b) The designation is composed of the group no. and serial numbering within a group.

e. g. 4.12:   Element no. 12 in group 4

Classification of groups:

Group 0:   All energy supply elements
Group 1, .2, .3 . . .:   Designation of the individual control chains (per cylinder, normally one group no.)

System for serial numbering:

.0:	Working elements
.1:	Control elements
.2, .4.: (even numbers)	All elements which have an influence on the advance movement of the working element concerned
.3, .5,: (odd numbers)	All elements which have an influence on the return movement
.01, .02 ...:	Elements between control element and working element, e. g. throttle valves.

Example 15:

Fig. 2/3 illustrates allocation of the designations to the elements.

Fig. 2/3

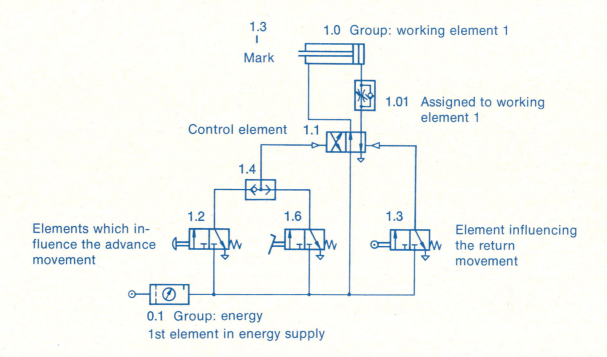

It must however be mentioned at this point that it is not always possible to provide unique allocation to the respective group or advance/return movement of a working element. In complex controls, overlaps usually exist, i. e. signals from one element act on various groups.

2.3.2.2 Designation using letters

This type of designation is used especially where circuit diagrams are being developed methodically. To some extent, this involves calculations and listings which can be performed more easily and more clearly when using letters.

Working elements are designated by capital letters, signal elements and limit switches by small letters.

In contrast to the preceding type of designation, the limit switches or signal elements are not assigned to the group which they influence but to the cylinder which acts on them. An allocation of this type is illustrated in Fig. 2/4.

Fig. 2/4

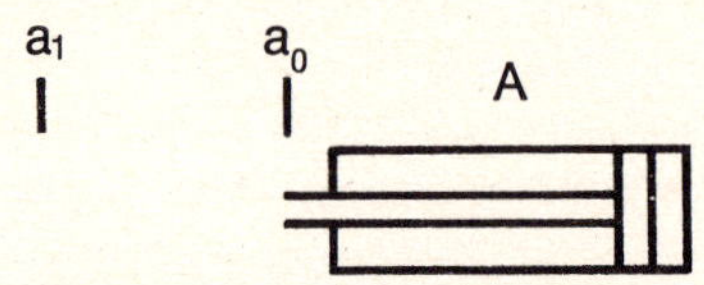

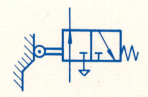

A, B, C, ...	Designation of working elements
$a_0, b_0, c_0 \ldots$	Designation of the limit switches which are actuated in the rear end positions of cylinders A, B, C, ...
$a_1, b_1, c_1, \ldots$	Designation of the limit switches which are actuated in the extended rod position of cylinders A, B, C, ...

It should also be noted that it is possible, similarly to electrical engineering practice, to use a combination of digits and letters when designating elements.

Threaded fittings and pipelines can be related to the designation of the elements by adding a suffix (e. g. 4.6 P, pressure connection of element 4.6).

2.3.3 Representation of devices

All devices should be shown in the circuit diagram in the control starting position. If this is not possible, or if this rule is not adhered to, an appropriate note must be made.

If valves have been drawn with normal position actuated, this must be indicated for example by an arrow, or in the case of a limit switch by drawing in the cam.

Definition of positions according to DIN 24 300

Normal position: Position assumed by the moving parts of the valve when the valve is not connected. (For valves with return means.)

Initial position: Position assumed by the moving parts after the valve has been installed in a system and the supply pressure has been connected, and it is the position in which the intended switching program commences.

Example 16:

Portrayal of a limit switch with closed normal position, drawn in actuated position (Fig. 2/5).

Fig. 2/5

2.3.4 Graphic symbols

In the completed circuit diagram, the same graphic symbols must be used as those provided on the elements in the completed installation.

2.3.5 Pipelines

Wherever possible, pipelines should be drawn as straight lines and without crossovers. Working lines are drawn as full lines, control lines as broken lines. It is simpler and clearer however, especially in complex controls, if the control lines are also drawn as full lines.

2.3.6 Pipeline designations

Pipelines may be given designations both in the circuit diagram and on the completed installation. A coded designation is recommended showing the connection and the destination. The connection code is made up of the number of the element and the connection (Fig. 2/6).

Fig. 2/6

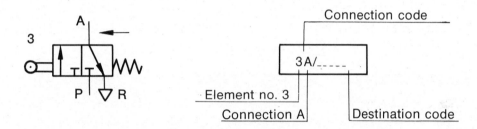

The destination code should specify where the line goes to, e. g. to element no. 12 and control connection X.

Complete connection code:

Fig. 2/7

on element 3 | 3A / 12 X | on element 12 | 12 X / 3A |

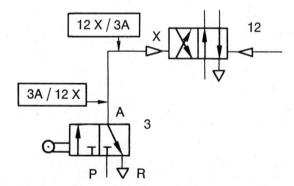

2.3.7 Additional information in the circuit diagram

Technical data for the equipment, setting values etc. may be included in the completed circuit diagram.

Furthermore, the sequence of movements should be described, in the form of a displacement-step diagram for instance; the incorporated boundary conditions should be listed; and the parts list for the working and control elements should be included.

2.3.8 Modified arrangements

If the circuit is contained in a control cabinet, the circuit diagram may be broken down into 3 parts.

a) Part 1 contains all elements which input signals to the control cabinet.

b) Part 2 contains the complete circuit contained in the control cabinet, including the operator controls and visual indicators fitted there.

c) Part 3 contains all elements to which the output signals from the control cabinet are applied.

The interconnections between the various parts may be designated by means of connection codes.

2.3.9 Summary

Examples of complete circuit diagrams are given in Part 4: Appendix.

- Circuit diagram layout as for control chain flowchart, signal flow from bottom to top wherever possible.
- Energy supply from bottom to top. Can be shown in simplified form.
- Spatial arrangement of the elements is ignored. Draw cylinders and directional control valves horizontal wherever possible.
- Designate all elements identically both in the circuit diagram and in the completed installation.
- Identify position of signals by a mark. If signals are issued in one direction only, arrow on the mark.
- Show devices in the initial position of the control. Identify actuated elements by a mark.
- Graphic symbols for the elements in the circuit diagram to be the same as on the devices in the completed installation.
- Drawn pipelines straight without crossovers wherever possible.
- If required, put in the pipeline designations.
- If required, enter technical data, setting values etc.
- Enter sequence of movements, boundary conditions and parts list.

2.4 Basic pneumatic circuits

In each branch of engineering the elements display certain peculiarities and fundamental characteristics. When designing functionable controls, it is essential that these be known. With regard to pneumatics, the differing construction of the elements must be considered such as the distinctive design characteristics of valves such as slide valves, poppet valves, and the resultant features such as behaviour on reserval (lingering or rapid), actuating force, flow behaviour (one or two directions), etc.
This knowledge of the characteristics of the elements to be used in pneumatic controls is, as already mentioned, a prerequisite.

There are many possibilities for the design of circuit diagrams, but there is one thing common to each – they can be put together from certain defined basic circuits. Consequently, a knowledge of these basic circuits is essential and these basic circuits point out possible uses and areas of application of the various elements.

In the following, the more important basic circuits will now be described by means of exercises and at the same time specific points will be raised to which particular attention must be given for the circuit in question.

Finally, it should be mentioned that negative controls, that is controls having valves which are reversed by pressure relief, will not be discussed here. The reason, quite simply, is that the authors are of the opinion that this type of control has now been replaced by better types and is no longer to be recommended.

© by FESTO DIDACTIC

2.4.1 Exercise on basic controls

2.4.1.1 The piston of a single-acting cylinder is to travel out when a button is operated and return at once to its end position when the button is released.

2.4.1.2 As exercise 2.4.1.1, except the single-acting cylinder is to be replaced by a double-acting cylinder.

2.4.1.3 The piston of a large-volume single-acting cylinder (large dia., large stroke length, large distance between cylinder and valve) is to travel out after actuating a valve and return to its end position after the valve has been released.

2.4.1.4 A double-acting cylinder is to be controlled by two valves 1.2 and 1.3 such that the piston travels out when valve 1.2 is actuated and remains in the forward end position after 1.2 has been released until the reverse signal for the return movement is input trough 1.3.

2.4.1.5 As exercise 2.4.1.4, except a single-acting cylinder is to be used in place of the double-acting cylinder and in all positions only valves with spring return are used.

2.4.1.6 After reaching the forward end position, the piston of a double-acting cylinder should reverse on its own, provided the valve (button) which initiates the forward movement is no longer actuated.

2.4.1.7 A double-acting cylinder is to be controlled such that, after being given the start signal, the piston performs oscillatory motion between the end positions until the reverse signal is input. The cylinder should then remain stationary in the rear end position.

2.4.1.8 The means should be provided for the piston rod of a double-acting cylinder to travel out-and-in in jogging operation. It must be possible for the piston to be stopped in any desired position by releasing the button and, as far as it is possible with air, be held in position.

2.4.1.9 The piston speed of a single-acting cylinder is to be adjustable for both forward and return movements and be regulated separately.

2.4.1.10 Exercise 2.4.1.6 should be extended such that the advance speed can be variably reduced and that the speed is increased on the return movement.

2.4.1.11 A single-acting cylinder is to be made to travel out by means of push buttons at two separate locations, and it is to return to the rear end position when the push buttons are released.

2.4.1.12 The piston rod of a double-acting cylinder is to travel out only when a push button is operated momentarily and an acknowledgment signal exists at the same time from a limit switch indicating a particular state within the system (e. g. magazine full). The return movement is controlled by a limit switch in the forward end position of the piston rod.

2.4.1.13 The piston rod of a double-acting cylinder is to travel out on being given a manual starting signal and be reversed in the forward end position. The return movement may only take place, however, if the max. pressure has built up in the end position of the cylinder.

2.4.1.14 After operating a manual button, the piston of a double-acting cylinder is to travel out, remain stationary in the forward end position for a certain adjustable period of time, and then return automatically.

2.4.1.15 The piston rod of a double-acting cylinder is to travel out when an air barrier is momentarily interrupted; its return movement is to be controlled by a proximity type of limit switch (reflex sensor) in the forward end position.

2.4.1.16 A double-acting cylinder is to be controlled in both directions by back-pressure nozzles. The sequence of movements is to be initiated and terminated by a switch. The initial position is always to be the rear end position.

2.4.1.17 By operating a button, the forward and return movements of a double-acting cylinder are to be controlled alternately.

2.4.2 Basic circuits using directional control valves
Exercises 2.4.1.1 – 2.4.1.8

2.4.2.1 Control of a single-acting cylinder
(Exercise 2.4.1.1)

Fig. 2/8

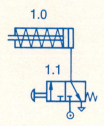

For this control function, a 3-way valve is required in order to exhaust the cylinder chamber on the return stroke as shown in Fig. 2/8.

2.4.2.2 Control of a double-acting cylinder
(Exercise 2.4.1.2)

Fig. 2/9

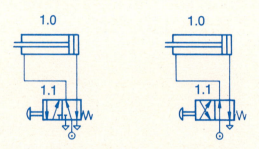

Either a 4-way valve or a 5-way valve can be used here (Fig. 2/9). When using a 5-way valve, it is possible to remove the exhaust air from the forward and return movements separately (e. g. speed regulation). As the 4-way function can also be obtained by combining two 3-way functions, it is also possible to use two 3-way valves for this control. With this type of switching, predetermined overlaps can be obtained.

2.4.2.3 Indirect control of a single-acting cylinder
(Exercise 2.4.1.3)

Fig. 2/10

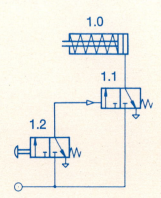

The type of control shown in Fig. 2/10 offers advantages especially with large-volume cylinders and long control lines. Owing to the indirect actuation of valve 1.1, signal element 1.2 can be kept small while the main valve has characteristic data corresponding to the cylinder size. The supply line from the control element, in this instance also called main valve, to the cylinder can therefore be kept very short, which means that the dead space and consequently the air consumption can be kept small, while the path signal element – control element is linked by a control line of small cross-sectional area.

This brings with it the additional advantage of manageable size, that is small external dimensions of the signal element, and also a reduction of the switching time.

2.4.2.4 Indirect control of a double-acting cylinder

(Exercise 2.4.1.4)

Fig. 2/11

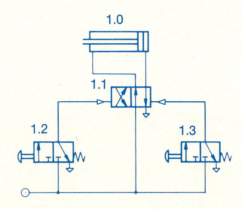

In this circuit (Fig. 2/11) there is no possibility of direct control. Valves 1.2 and 1.3 would indeed be capable of moving the piston forwards and back if connected directly to the cylinder, but the piston would be depressurized in each end position and would thus not be fixed.

Thus, in general one must differentiate between

– direct and
– indirect control of a cylinder.

Direct control may be selected only if the cylinder volume is not too large and particularly if the process to be influenced can be controlled by **one** signal element.

It should still be mentioned that this type of control is also known as impulse control. The control element in this case is a valve which is reversed at both ends by applying pressure. An impulse is sufficient here to switch the valve over and it can by retained in the new position by restraining forces such as friction. This type of valve is also known as "impulse valve" (valve with memory characteristics).

2.4.2.5 Indirect control of a single-acting cylinder using a holding control

(Exercise 2.4.1.5)

In order to solve this problem, the term holding control must be introduced. Whereas the impulse valve in Fig. 2/11 causes an input signal to be stored, i. e. an output signal is still present after the input signal has been removed (friction in slide valve), in the spring return valve the output falls off when the input signal is removed. Thus, if a memory function is required, a special circuit arrangement has to be devised. A suitable arrangement is the self-holding circuit.

Fig. 2/12

In the circuit shown in Fig. 2/12, the signal from T_2 flows through the nonoperated valve T_1 and switches over valve V_1, which means that air is available at the outlet. If T_2 is now released, V_1 maintains its condition through V_2 and T_1 even in the operated position. If the self-holding condition is to be interrupted, T_1 must be operated. If T_1 and T_2, are operated together, the flow passage to the control inlet of V_1 is blocked, the arrangement remains in its initial position or T_1 (in this example "OFF") dominates. If T_1 is connected between V_1 and V_2, signal T_2 (i. e. "ON") dominates.

Fig. 2/13

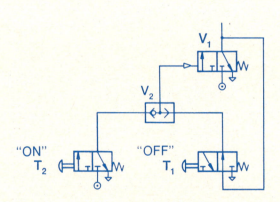

Normally, it would now be possible to connect the controlled cylinder to the outlet of V_1, but this is not altogether advisable. Since the pressure build-up in the cylinder does not take place at once, and since the pressure is also not constant over the stroke, but on the other hand V_1 requires a certain minimum pressure for reversal, the self-holding function may not be effective if T_2 is released too early.

Furthermore, when T_2 is operated the control side of V_1 is actually exhausted, that is V_1 switches over and the cylinder is exhausted, but if T_1 is released prematurely the residual pressure in the not completely exhausted cylinder will be sufficient to switch V_1 back again. These disadvantages can be avoided if another valve 1.1 is connected to the outlet of V_1. This then also results in the solution to exercise 2.4.1.5 in Fig. 2/14. (In this case, drawn for dominant "OFF"!)

Fig. 2/14

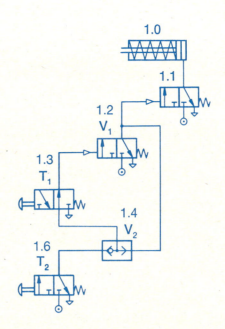

One point should still be mentioned:
When building up a self-holding circuit, it must be ensured that the self-holding action is provided through an intermediate valve and not directly on the valve which is to be held as the latter could lead to difficulties when exhausting (Also applies to elements with overlap.)

Incidentally, the requirement to use only valves with spring return can, for example be imposed for reasons of safety if a definite starting position is to be assumed in the event of a power failure.

2.4.2.6 Automatic return control of a double-acting cylinder using a limit switch

(Exercise 2.4.1.6)

Fig. 2/15

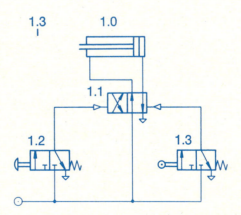

One characteristic of the impulse valve is illustrated by the circuit in Fig. 2/15. If too long a signal is given to the impulse valve through valve 1.2, the cylinder in the forward end position does in fact move to valve 1.3 and provides the signal for the return. However, this signal cannot be effective as the signal from 1.2 still exists on the opposite side of 1.1. (With equal pressure and equal pilot spool areas, valve 1.2 is held in its present position by the friction when there is force equilibrium.) Reversal can take place only when 1.2 is released. Thus we have here a control element with dominant behaviour with respect to the signal first received.

2.4.2.7 Continuous reciprocation of a double-acting cylinder with means of switching off

(Exercise 2.4.1.7)

Fig. 2/16

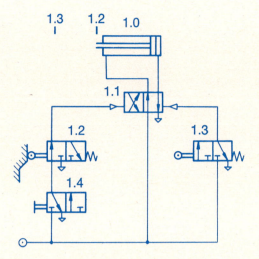

As valve 1.2 is actuated in the rear end position and thereby provides the signal for the forward movement, switching off can be effected simply by removing the air from 1.2 (Fig. 2/16). Otherwise, as already described in 2.3, a system is always drawn in the initial position. Since in this particular case 1.2 is actuated in the initial position of the system, it should also be drawn as such in the circuit diagram (with note!).

2.4.2.8 Stopping and fixing a double-acting cylinder in intermediate positions

(Exercise 2.4.1.8)

Owing to the compressibility of compressed air, it is not possible to effect accurate stopping of a cylinder in intermediate positions. Nonetheless, there are cases in practice where small load deviations occur during the stationary condition and where the achievable accuracy suffices. Fig. 2/17 shows a circuit for this case.

Fig. 2/17

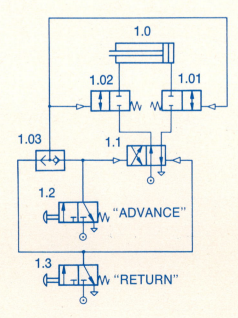

© by FESTO DIDACTIC

In this circuit, both supply lines to the cylinder are opened through the 2-way valves 1.01 and 1.02 when buttons 1.2 and 1.3 are actuated and this enables the cylinder to move. When the respective signal element is released, the supply and exhaust lines are blocked, the piston moves until a state of equilibrium is reached and then remains stationary, held by the residual pressure.
Note: Valves 1.01 and 1.02 must be such that air can flow in both directions.

A further solution is provided by using a 3-position valve with fixed neutral position and reversal from both sides by applying pressure. Fig. 2/18 shows this circuit.

Fig. 2/18

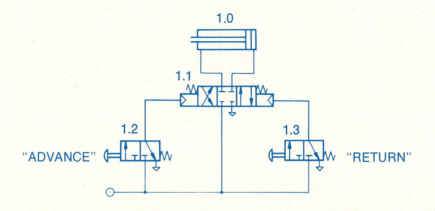

2.4.2.9 Summary

To sum up, here are the more important points from this chapter.

Areas of application and use of directional controls valves with

2-way function:

— For pure shut-off problems

3-way function:

— Control of single-acting cylinders
— Control of valves, reversed by applying pressure
— General: Anywhere where a signal is required for initiating a procedure and must be exhausted through the valve used.

4-way function:

— For controlling double-acting cylinders and as shuttle valve for signal linkages.

5-way function:

— As for 4-way function, but equipped with 2 exhaust ports (one for each working line). The exhaust can be utilized separately (e. g. speed regulation).

Distinction:

— Direct control
— Indirect control

Direct control:

— Where **one** signal is sufficient for controlling and where it is not required that large-volume cylinders be controlled.

Indirect control:
– Where there are several signals and where control elements and signal elements cannot be combined.

Distinction:
– Impulse control
– Holding control

Impulse control:
– Normally used in pneumatics as the impulse valve represents a low-cost and high-value control element.

Holding control:
– Only where specific requirements are imposed on the system (e. g. safety, definite normal position, etc.). Expensive, compared to impulse control.

2.4.3 Circuits for speed regulation on cylinders
(Exercises 2.4.1.9 – 2.4.1.10)

One must differentiate here between

– reducing the speed
– increasing the speed

Reducing the speed:
This is achieved by using throttle valves. If only one direction of motion is to be influenced, a check valve is connected in parallel to the throttle valve.

Depending on the type of equipment, there are three possibilities which can be considered.

a) Constant, non-adjustable restriction. Fig. 2/19 shows the symbolic representation.

Fig. 2/19

b) Throttle constant over the stroke, but manually adjustable (Fig. 2/20).

Fig. 2/20

c) Throttle continuously adjustable over the stroke (by operation of the throttle valve roller) in Fig. 2/21.

Fig. 2/21

© by FESTO DIDACTIC

Regarding installation of throttle valves, there are two possibilities:
- supply air throttling
- exhaust air throttling

It is assumed that the advantages and disadvantages of these are known and for this reason they will not be discussed further here.

Increasing the speed:

Purely from a control engineering point of view, the speed can be increased only by fitting a quick exhaust valve.
If this in itself is not sufficient, other methods must be applied such as changing the port areas, different layout and selection of the elements, etc.

2.4.3.1 Speed regulation on a single-acting cylinder

Advance:

With a single-acting cylinder, throttling of the advance movement can be effected only by supply air throttling (exhaust is not contained), and it is quite impossible to increase the speed by means of a quick exhaust valve.
Fig. 2/22 shows supply air throttling for the advance movement.

Fig. 2/22

Return:

There is no option but to provide exhaust air throttling here.
Fig. 2/23 shows exhaust air throttling for the return movement.

Fig. 2/23

Fig. 2/24 illustrates the use of a quick exhaust valve to increase the return speed.

Fig. 2/24

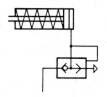

Advance and return:

Reduction of advance and return speeds, separately adjustable, can be seen in Fig. 2/25.

Fig. 2/25

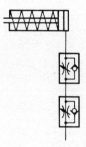

Fig. 2/26 shows one means of obtaining non-separately adjustable reduction of advance and return speeds.

Fig. 2/26

A non-adjustable reduction of advance and return speeds is illustrated in Fig. 2/27.

Fig. 2/27

2.4.3.2 Speed regulation on a double-acting cylinder

It is here possible to apply supply air and exhaust air throttling to advance and return movements. At the same time, it is possible to regulate in both directions based on distance travelled by means of a roller-actuated throttle valve. A quick exhaust valve can also effect speed increase in both directions. Combinations are possible.

Fig. 2/28 shows a separately adjustable supply air throttling for advance and return.

Fig. 2/28

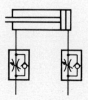

Exhaust air throttling for advance and return, separately adjustable, is shown in Fig. 2/29.

Fig. 2/29

If a 5-way valve is used to control the cylinder, the throttle valve can also be placed in the exhaust port of the valve. The rebound effect is thus avoided (Fig. 2/30).

Fig. 2/30

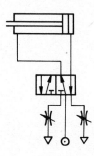

Fig. 2/31 shows the exhaust air throttled travel-dependent speed regulation for the return movement.

Fig. 2/31

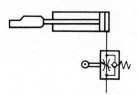

A speed increase arrangement for the advance movement by means of a quick exhaust valve is shown in Fig. 2/32.

Fig. 2/32

2.4.3.3 For exercise 2.4.1.9, the given considerations give rise to a solution according to Fig. 2/33.

Fig. 2/33

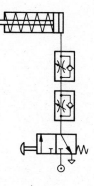

73

2.4.3.4 Circuit diagram for exercise 4.1.10 (Fig. 2/34)

Fig. 2/34

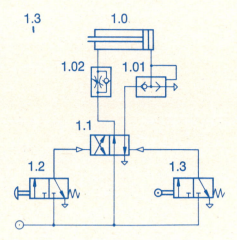

An exhaust air throttling has been selected here.

2.4.4 Circuits with shuttle and two-pressure valves
(Exercise 2.4.1.11 and 2.4.1.12)

In general terms, the areas of application of these valves can be described in the following manner:

2.4.4.1 Shuttle valve

Where the same operation is to be initiated by two signals, i. e. where signals are to be brought together. The shuttle valve is frequently termed "pneumatic OR element" because it has the basic logic function of "OR", i. e. a signal is given at the output if a signal is present at either one input or the other or at both inputs.

2.4.4.1.1 Controlling a single-acting cylinder from two different points
(Exercise 2.4.1.11)

Fig. 2/35

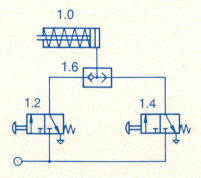

Without the shuttle valve in the circuit shown in Fig. 2/35, the air would escape through the exhaust of the other non-operated valve when 1.2 or 1.4 is operated.

If several signals are to be taken to one outlet, it is necessary to connect shuttle valves in series (as each valve has only two inlets).

Example 17:

It is required that 4 signals $e_1 \ldots e_4$ initiate the same operation, that is they are all to be linked to one outlet a. Two possible means of achieving this are shown in Fig. 2/36.

Fig. 2/36

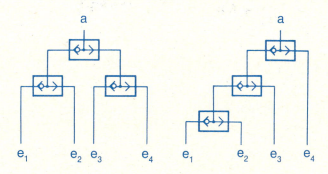

Required number of valves n_v for one outlet a:

$$n_v = e_n - 1$$

e_n: Number of signals to be grouped.

General: In pneumatics, a shuttle valve is always required where several signals are connected to one outlet. It is not required, however, if one signal is connected to several outlets.

Fig. 2/37

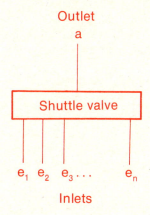

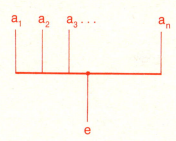

2.4.4.2 Two-pressure valve

An operation may be performed only when 2 signals are received simultaneously.

The two-pressure valve is also termed "Pneumatic AND element" because it has the basic logic function of "AND", i. e. a signal is given at the outlet only if both inlet signals are present.

Example 18:

The piston rod of a single-acting cylinder may travel out only if two 3-way valves are operated simultaneously. (Fig. 2/38).

Fig. 2/38

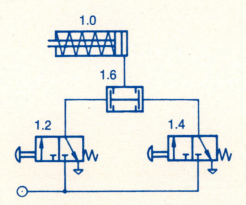

Here too it is possible to group several signals by connecting several two-pressure valves in series.

Example 19:

An operation may be initiated only if 5 signals $e_1 \ldots e_5$ are present. The number of two-pressure valves required works out at

$\underline{n_v = e_n - 1 = 5 - 1 = 4}$

Fig. 2/39 shows the possible arrangements of the two-pressure valves.

Fig. 2/39

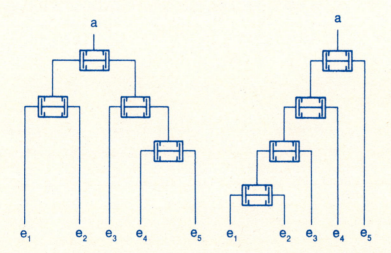

2.4.4.2.1 Control of a double-acting cylinder by means of two valves and one two-pressure valve
(Exercise 2.4.1.12)

Fig. 2/40

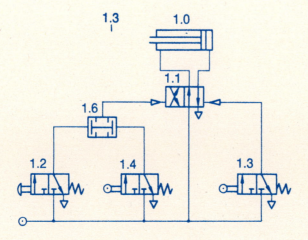

2.4.5 Pressure operated controls

(Exercise 2.4.1.13)

The sequence valve was developed in pneumatics for the purpose of controlling certain operations and factors by means of pressure. Owing to particular design features and the absence of an exhaust, it can in practice be used only in conjunction with a directional control valve.

Fig. 2/41

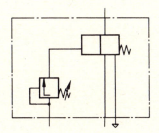

This unit is shown symbolically in Fig. 2/41. Since it rather complicated to draw this sequence valve in this way in practice, however, the form of representation shown in Fig. 2/42 is used in the drawings below. In this graphic symbol, the pressure relief is also shown. (This symbol is not yet standardized.)

Fig. 2/42

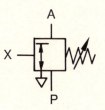

There are two possible solutions to the problem given in exercise 2.4.1.13.

2.4.5.1 Pressure controlled reversal with mechanical end-position checking using limit switches (Fig. 2/43)

Fig. 2/43

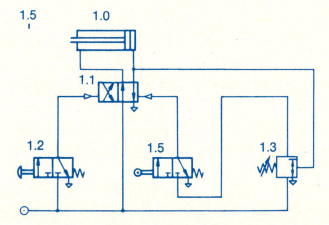

The sequence valve 1.3 should be set such that the switching point is just less than the prevailing working pressure.

As the maximum pressure can build up only when the piston is stationary in the cylinder, a signal can be given through the sequence valve only when the piston is stationary, a condition which arises normally only when the end position has been reached. To make certain that the end position has in fact been reached, limit switch 1.5 is also interrogated.

2.4.5.2 Pressure controlled reversal without mechanical checking of the end position (Fig. 2/44)

Fig. 2/44

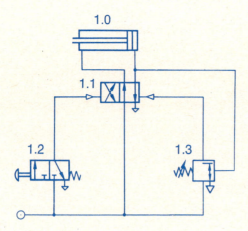

In this type of circuit, however, reversal does also occur if the piston is stopped in any intermediate position, that is where the maximum pressure can build up. This type of control should therefore be used only where the demands with respect to certainty of operation are not too great or where there is no possibility of using limit switches or alternatively where a reversal is required when a specific opposing force arises.

Furthermore, it is important with pressure operated controls to note the throttling positions and to connect the throttle at the inlet or the outlet according to the desired function.

2.4.6 Circuits with time behaviour

(Exercise 2.4.1.14)

Pneumatic time elements can be formed very simply from combinations of directional control valves, throttle relief valves, and volumes. Regardless of whether this combination has been formed by connecting single elements or by using a self-contained unit, the symbol for the time element is made up of the various functions required.

The more important and most common time elements are shown below.

It should again be mentioned that, in pneumatic timing circuits particularly, the construction of the elements used may not be disregarded since, for example, a poppet valve has quite different switching characteristics from a slide valve. (Rapid switch-over, counter-force, switching pressure).

One distinction should still be noted. This does not concern the circuit as such but rather the application group. It is quite possible for one circuit to be classified in both groups.

1. Time circuits for defined time-dependent reversal
2. Time circuits for pulse shaping

2.4.6.1 Time circuits for defined time-dependent reversal

Start-delayed time behaviour: This behaviour is illustrated in Fig. 2/45.

Fig. 2/45

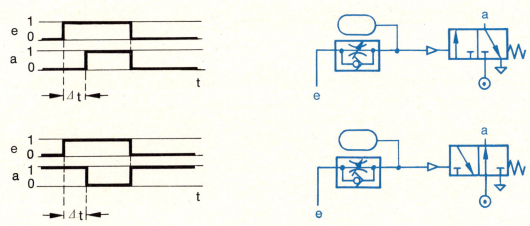

Falloff-delayed time behaviour: This behaviour is illustrated in Fig. 2/46.

Fig. 2/46

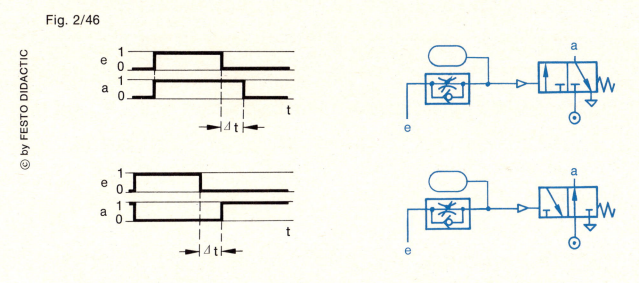

A **start- and falloff-delayed time behaviour** is shown in Fig. 2/47 (separately adjustable).

Fig. 2/47

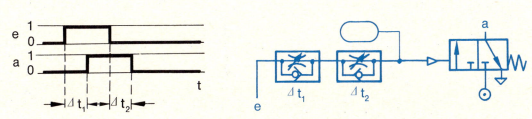

2.4.6.2 Time circuits for pulse shaping

The timing accuracy expected in this type of circuit is usually not too great. It is, however, important that the pulses are shaped accurately and cleanly.

Pulse shortening: (Fig. 2/48)

Fig. 2/48

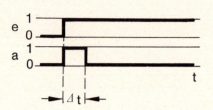
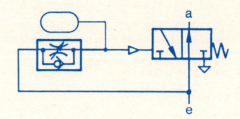

Pulse stretching: (Fig. 2/49, see also Fig. 2/46)

Fig. 2/49

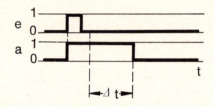
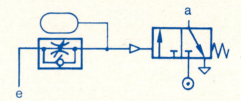

2.4.6.3 Exercise 2.4.1.14

This exercise can be solved in two different ways.

a) Time-dependent reversal of the piston after reaching and checking the end position (by means of limit switch, Fig. 2/50).

Fig. 2/50

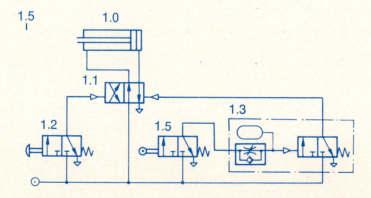

The time commences after the end position has been reached, 1.1 reverses at the end of the time set on throttle 1.3 and the piston returns.

b) Time-dependent reversal without checking of end position.

Fig. 2/51

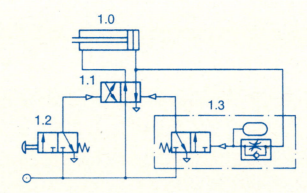

This type of circuit shown in Fig. 2/49 has the advantage that it works without limit switch, albeit at the expense of operational reliability. The time element is pressurized in this case as soon as 1.1 reverses, i. e. at the beginning of the piston outward stroke. This means that the pure dwell time of the piston rod in its forward end position is dependent on the duration of the outward stroke which can, as is well known, vary tremendously. Furthermore, the pressure build-up behind the piston is not constant, depending on the load applied, and this thus represents an additional source of error in the circuit. It must also be considered that, if the piston stops in an intermediate position, the standstill time extends the total time and therefore in an extreme case the end position may not even be reached.

2.4.7 Basic circuits with contactless signal transmitters
(reflex sensor, air gate, and back-pressure nozzle)

(Exércises 2.4.1.15 and 2.4.1.16)

It would be beyond the scope of this work to discuss all contactless sensing systems in pneumatics, or rather low-pressure pneumatics to use the correct term. This subject will be dealt with in the seminar on low-pressure engineering.

Only the basic circuits for 3 types will be demonstrated:

— reflex sensor
— air gate
— back-pressure nozzle

As no universally valid symbols have been established for these systems, the representation must first be discussed.

Reflex sensor:

This is an element from which air is emitted through an annular nozzle, and thus must be supplied with air, and which puts out a signal through a centrically arranged channel. 2 connections are required. The symbol is shown in Fig. 2/52.

Fig. 2/52

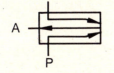

A: Signal output
P: Connection for supply line

Air gate:

This comprises an emitter nozzle and a supplied collector nozzle.

Fig. 2/53

Back-pressure nozzle:

This is only to a limited extent a contactless switching element. With the back-pressure nozzle, the outlet opening must be closed completely if full effectiveness is to be obtained. The actuating force is very low, though (area of opening x pressure applied). One advantage of the back-pressure nozzle is that it can be supplied with normal pressure and also that the output signal is at normal pressure level.

Fig. 2/54

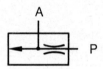

Since the output signals from the air gate and the reflex sensor are low-pressure signals, it is necessary to incorporate amplifiers. As these amplifiers must be supplied with low pressure in order to attain the high level of amplification, the general amplifier symbol is also provided with a supply pressure connection.

Fig. 2/55 shows the symbols for the amplifiers with and without supply.

Fig. 2/55

Since a directional control valve is operated by the amplified signal, the symbol may also be drawn directly on the directional control valve as shown in Fig. 2/56.

Fig. 2/56

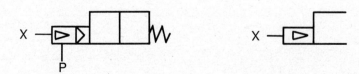

Outward travel of a double-acting cylinder by means of an air gate and return controlled by a reflex sensor.

(Exercise 2.4.1.15)

Fig. 2/57

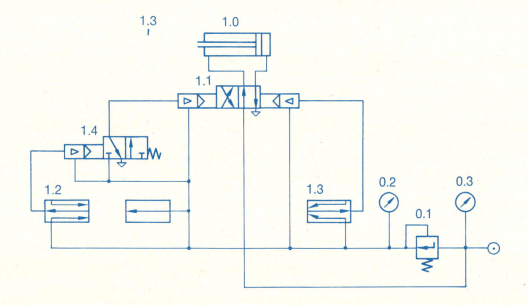

It is necessary to connect valve 1.4 between 1.2 and 1.1 because the air gate outputs a signal in the uninterrupted state whereas this signal is required only when the air gate is interrupted. Thus, a valve with open normal position must be used for reversal.

If the signal from the reflex sensor is also provided through an intermediate valve, the circuit shown in Fig. 2/58 applies. This arrangement has the advantage that the signals from 1.2 and 1.3 are available as normal pressure signals after valves 1.4 and 1.5 and can thus be utilized directly in the control.

Fig. 2/58

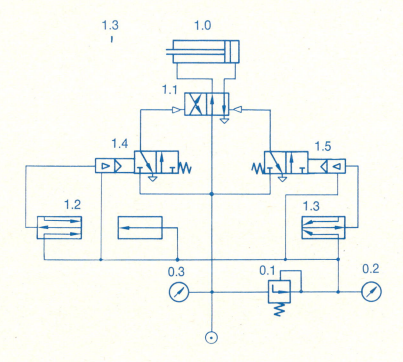

Controlling the advance and return movements of a double-acting cylinder by means of back-pressure nozzles

(Exercise 2.4.1.16)

Fig. 2/59

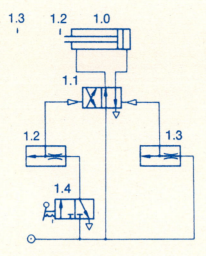

The cylinder movement can be initiated and terminated by means of switch 1.4 in Fig. 2/59. The advantage of the back-pressure nozzles is that they can be supplied with air from the normal network and no amplification is required. The outlet openings must, however, be fully closed.

2.4.8 Various basic circuits

2.4.8.1 Alternating controls

(Exercise 2.4.1.17)

The characteristic of an alternating control is that the output reverses with each arriving input signal (emitted from one signal transmitter only).

Fig. 2/60 shows the relationship between output signal a and input signal e.

Fig. 2/60

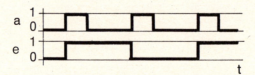

Every other input signal thus causes the output signal to fall back to its original position. This circuit is therefore also known as a binary reducing stage or a reducing circuit.

These alternating or reducing circuits are very important in control engineering because they are indispensable for counting circuits, register circuits etc.

There are various possible solutions for the exercise on hand, the alternate outward and return movement of a cylinder.

2.4.8.1.1 Alternating circuit with latching by means of limit switches

If limit switches can be mounted at the end positions of the cylinder, and if these limit switch signals are not required for other purposes also, then the circuit shown in Fig. 2/61 can be used.

Fig. 2/61   Fig. 2/62

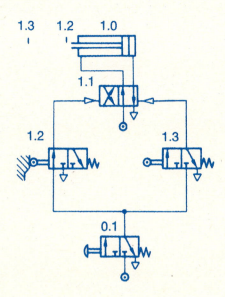
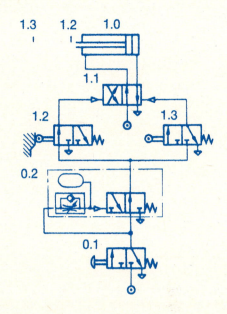

It does however have the disadvantage that it will be reversed immediately if the input signal is applied for too great a length of time in the forward end position. If this is to be prevented, the signal duration of the input signal must be less than the stroke time of the piston. This condition can be obtained by, for example, a circuit for reducing the impulse length of the input signal as shown in Fig. 2/62.

2.4.8.1.2 Alternating circuit with latching by means of impulse valve and shuttle valve

Fig. 2/63

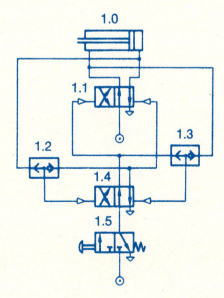

In the circuit shown in Fig. 2/63, use is made of the fact that in the case of an impulse valve the signal which first arrives dominates. In the position drawn, valve 1.1 is reversed when a signal is given through valve 1.5 and 1.4 is held in the initial position by the signal via 1.3; 1.4 can be reversed by the output signal from 1.1 only when 1.5 is released thereby opening the path for the return signal.

With this type of control, disturbances can however arise if undefined conditions prevail due to unfavourable build-up or fall in the cylinder, unsymmetrical construction of the circuit or due to poor impulse transmission through the signal element. Some improvement is obtained if the cylinder is not pressurized directly from the output of the alternating control but from an additional impulse valve. (Fig. 2/64).

Fig. 2/64

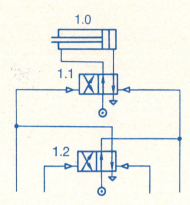

Controls which are functionally reliable and free from disturbances can be built in accordance with the cicuits shown in Fig. 2/65 and 2/66.

2.4.8.1.3 Alternating circuit with latching by means of impulse valve and pressurized 3-way valve with spring return. (Fig. 2/65)

Fig. 2/65

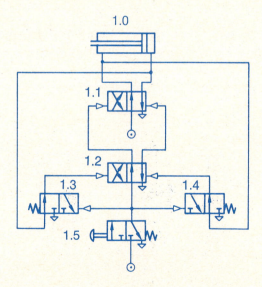

When the signal element 1.5 is operated, the feedback from output 1.1 via valves 1.3 and 1.4 is interrupted and re-established only after 1.5 has been released, i. e. 1.2 is reversed only at this moment and the path for the feedback signal opened.

2.4.8.1.4 Alternating circuit with latching by means of impulse valve and two-pressure valve. (Fig. 2/66)

Fig. 2/66

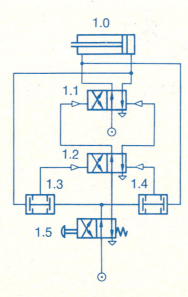

A disadvantage of this circuit is that a 4-way valve must be used as signal transmitter. Also in this case, reversal of 1.2 can take place only when 1.5 is released thus causing the two-pressure valve required at this instant to be pressurized from both sides.

2.4.8.1.5 Valves with alternating behaviour

Because all these alternating circuits are very elaborate and expensive, valves have been developed which make this alternating behaviour possible.
Unfortunately, there are not yet any standardized symbols for these valves either, and thus a definition must be made. The symbol shown in Fig. 2/67 will be used in the following text to represent the alternating behaviour.

Fig. 2/67

A circuit for exercise 2.4.1.17 could then appear as shown in Fig. 2/68:

Fig. 2/68

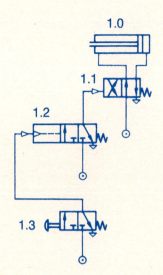

2.4.8.2 Circuit for reciprocating movements of a cylinder without limit switch

If a cylinder is to perform reciprocating movements without using limit switches, time elements can be used to effect reversal. This has the advantage that frequency and stroke can be changed time-dependently. One possible circuit is shown in Fig. 2/69.

Fig. 2/69

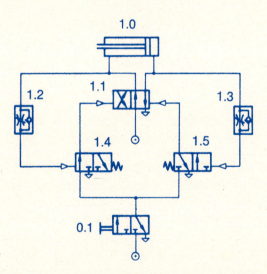

© by FESTO DIDACTIC

In this circuit, the piston remains stationary in one of the end positions after switching off by means of 0.1. The end position assumed depends on the direction of movement at the moment of switching off.
If a defined initial position is to be assumed, air may be removed from only one of the valves 1.4 or 1.5, depending on the position desired.
Circuit 2/70 shows a more simple construction. The switching arrangement is known as a multivibrator valve unit.

Fig. 2/70

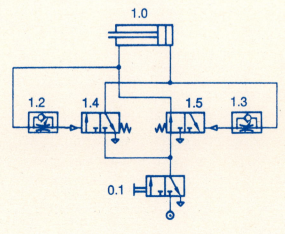

2.4.9 Circuits for signal suppression and signal elimination

A continually recurring problem in control engineering is the suppression or elimination of nonrequired signals. (e. g. signals from limit switches and manual controls which are still operated when the opposing signal occurs and then do not permit this opposing signal to be effective.)

In pneumatics, there are various possibilities for eleminating these signals, and one must distinguish between two main groups.

1. the signal which is still applied is overridden by a stronger signal (signal suppression)
2. the applied signal is cut out (signal elimination).

2.4.9.1 Circuits for signal suppression

As already mentioned, the signal which is still applied is overridden by a stronger signal in this type of circuit. In hardware terms, this can be accomplished either by means of a directional control valve with differential pressure operation (Fig. 2/71) or by means of an impulse valve with a regulator connected on one of the control sides (Fig. 2/72). (Signal a is overridden by signal b!)

Fig. 2/71

Fig. 2/72

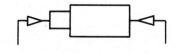

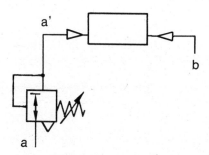

In terms of function, both possibilities are equally good. To ensure reliable operation, however, it is necessary for, in Fig. 2/71, signal a and signal b to have the same pressure level, or in Fig. 2/72 for the chosen pressure difference between a' and b to be of sufficient magnitude.

It should still be noted that if signal a is applied continuously, signal a is again effective after removing signal b and a reversal occurs.

2.4.9.2 Circuits for signal elimination

In these circuits, it is possible to remove signals either mechanically or by means of suitable circuitry.

1. Mechanical signal elimination

a) Elimination by means of short-impulse transmitters.
   The symbol for this valve is shown again in Fig. 2/73.

Fig. 2/73

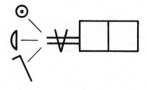

It is composed of a directional control valve, the symbol for the over-centre function, and the respective type of control.

When using this valve, the following points must be observed:
- the operating reliability depends to a great extent on the speed of actuation (max. 0.1 – 0.15 m/sec., to still obtain a sufficiently long signal).
- In the middle part of the actuation stroke the valve is operated, but not in the end positions. This means that the actuator must be operated as far as the stop as otherwise a continuous signal will exist. (Firm mounting of limit switches).
  As a result, however, the signal no longer exists after complete actuation, and can thus also not be used for further control operations or for checks of the prevailing position of the whole system.
- When used as a limit switch, the switching point is not exactly in the end position (approx. 4–5 mm ahead of this).

b) Elimination by means of idle return roller
   If the signal to be eliminated is provided by a limit switch, a valve operated by an idle return roller can be used. (Symbol in Fig. 2/74).

Fig. 2/74

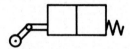

In application, the following points must be obversed:
- the signal cannot be produced in the end position as the valve must be passed over completely. The position of the switching point depends on the detail design of the element, on the rate of traverse, and on the length of the control cam.
- Since the valve is again released in the end position, there are no means here either for making further use of the signal for subsequent operations or for checking purposes (see also under [a] above).

2. Circuits designed for signal elimination

c) One possibility is to use the circuit for pulse shortening described in Section 2.4.6.2 (Fig. 2/75).

Fig. 2/75

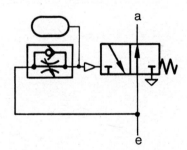

These circuits are very reliable in operation, but when used in elaborate controls they are complicated and expensive.

In addition, they can be used solely for signal elimination and do not enable latching against multiple operation to be incorporated.
This can be seen clearly from Fig. 2/76.

Fig. 2/76

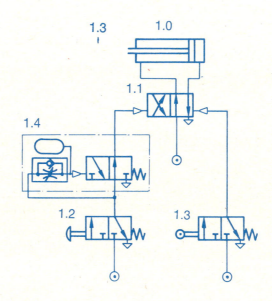

In this circuit, the signal from 1.2 is always transformed to a pulse and signal 1.3 can thus be effective. If 1.2 is operated again during the return movement of the piston rod, however, a further reversal takes place.

d) Signal elimination by means of reversing valve.

This method is certainly the most frequently used in practice. Provided the layout is correct, it functions very reliably and has the additional advantage that frequently several signals can be grouped together for elimination and the expenditure thus remains relatively low.

The basic idea is to allow the signal to be effective only at the time when it is needed. This can be accomplished either by blocking the signal behind the signal element by means of a valve, or by supplying energy to the signal element only when the signal is required. An impulse valve is normally used to effect reversal.

Difficulties in practice usually concern the selection of the reversing signals for the reversing valve used in this type of elimination. This will be discussed in more detail elsewhere.

2.4.9.3 Summary

Fig. 2/77 contains a brief summary of the means of providing signal suppression and signal elimination.

Fig. 2/77

Signal suppression:

Differential pressure operation

Pressure regulator

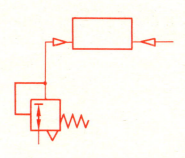

Signal elimination:

by over-centre device

by roller plunger with idle return

pulse shortening

reversing valve

2.5 Methods for constructing a circuit diagram

Basically there are two primary methods for constructing circuit diagrams.

1. The so-called "intuitive" methods, also frequently termed conventional or trial-and-error methods.
2. The methodical design of a circuit diagram in accordance with prescribed rules and instructions.

In the following text, the first group will include all types of circuit diagrams which are constructed mainly on the basis of experience. This will not however necessarily exclude a methodical approach which may indeed frequently also be essential, but in this case the individual will have a greater influence than the methodical development.

The second type will include all types in which a precisely defined method is applied, that is the personal influence of the designer is less.

Whereas much experience and intuition is required in the first case, and also a great deal of time above all with complicated circuits, designing circuit diagrams of the second category requires methodical working and a certain amount of basic theoretical knowledge.

Regardless of which method is used in developing the circuit diagram, the object is to end up with a properly functioning and reliably operating control. Whereas previously much emphasis was placed on the least expensive solution, more importance is now attached to operational reliability and ease of maintenance and hence to the clear layout.

This necessarily leads to increased methodical design of circuit diagrams. In such cases, the control is constructed at all times in accordance with the given procedure, independent of personal influences from the designer such as ability, frame of mind, mood and so on, and is thus clear to other people and could be completed by them who may need to become involved with this control. In most cases, however, more devices will be required in such a control than in a circuit devised by the intuitive method.

This additional material requirement will, however, usually be rapidly compensated for by a time-saving at the project stage and also later in terms of maintenance. Generally, it must be ensured that the time spent in project design and particularly in simplifying the circuit is then a reasonable proportion of the overall effort.

At this point, one particular point is again emphasized: regardless of which method and which technique is used to produce a circuit diagram, sound fundamental knowledge of the devices concerned and knowledge of the switching possibilities and characteristics of the elements used are basic requirements.

It will now be shown in the following how simple pneumatic circuits can be developed and constructed while still making use of simpler methods.

It is important to pay attention to the type of control involved, and this can usually be seen from the problem definition. A circuit diagram for a memory control will be built up differently from, for example, a circuit diagram for a coordinated motion control.

Since program controls are the most frequently encountered controls in pneumatics, and these for the most part are co-ordinated motion controls, this type of control will be treated in some detail.

Finally, the reader is reminded again that the method for dealing with a control problem is given in Part 1, Section 8 and that the points listed must be dealt with and clarified before producing the circuit diagram.

2.6 Constructing the circuit diagram for coordinated motion controls

If the motion diagram and auxiliarly conditions have been clearly defined, drawing of the circuit diagram can commence. Representation and drawing layout were described in detail in Section 2.3. The circuit is now built up in accordance with this pattern.

This design, and hence also the basic pattern for constructing the circuit diagram, depends on the type of signal cut-out used.

Where simpler types of controls are concerned, and particularly where the disadvantages of cutting out signals by means of idle return rollers can be accepted, the use of valves controlled by idle return rollers can be justified.

In all other cases, signal cut-out by means of reversing valves should be incorporated.

Incorporation of these valves can be effected in different ways. First, it will be shown how this type of cut-out can be incorporated by the intuitive method, and then one of the available methods will be applied which always ensures positive and reliable cut-out by means of reversing valves, even where the control is elaborate.

This methodical approach towards producing a circuit diagram is also called the "Cascade Method".

The advantages and disadvantages will be discussed in detail later in the text. At this point, suffice it to say that this method for constructing a circuit diagram is certainly the easiest to learn for controls where signal cut-out is effected by means of reversing valves. The same basic thinking is also behind the design of a shift register circuit.

Another point to be observed is in the inclusion of auxiliary conditions in a control. It is expedient to consider and include these conditions only when the basic circuit for a function has been completed. These requirements should then be incorporated singly, i. e. the circuit diagram should be expanded step by step. This is the only way to ensure that the circuit retains its overall clarity, even where elaborate controls are concerned.

Exercises and examples will be given which should assist in properly grasping the methods and tips listed here. At the same time, these examples will allow some important circuitry tricks to be learned.

Naturally, most control engineering problems have several solutions each of which is frequently as good as the next. For the sake of clarity, only those solutions will be given in the following which are most important to know and which are considered most suitable from the point of view of learning success.

2.6.1 Exercise 6.1 Package transfer

Problem:

Packages arriving on a roller-conveyor are lifted by a pneumatic cylinder and pushed onto another conveyor by a second cylinder. Cylinder B may then perform a return stroke only after cylinder A has reached the rear end position. The start signal should be provided by means of a manual button, each signal initiating one cycle.

Positional sketch and determination of working elements:

Fig. 2/78

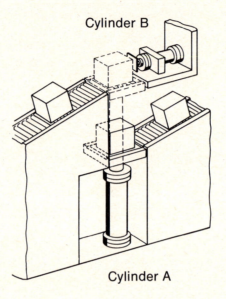

Displacement-step diagram:

Fig. 2/79

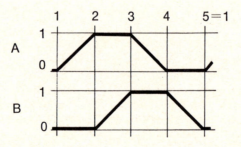

2.6.1.1 Circuit design for exercise 6.1

As already mentioned, the procedure adopted in designing a circuit diagram depends on the type of signal cut-out.

The circuit design is simplest if signal cut-out is arranged by means of idle return rollers.

The following procedure is then recommended in building up the circuit diagram:

a) Draw the working elements.

b) Draw the associated final control elements.

c) Draw the required signal elements without symbol. If impulse valves are used as final control elements, 2 driving signals and hence 2 signal elements are required initially per impulse valve.

d) Draw in the energy supply.

e) Connect control lines.

f) Number the elements.

g) Transpose the motion diagram into the circuit diagram (programming of the circuit by positional allocation of the limit switches to the respective working element).

h) Check where signal cut-outs are required. This check can be made in the motion and control diagram.

i) Draw in the actuating controls.

k) Where applicable, incorporate the auxiliary conditions.

By adopting the procedure according to items a) – g), the circuit shown in Fig. 2/80 results for exercise 6.1.

Fig. 2/80

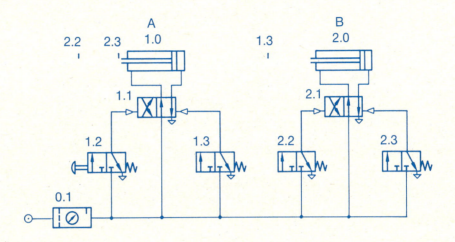

A check can be made in the motion and control diagram (Fig. 2/77) to determine whether and where a signal cut-out is required.

In general, the control diagram is drawn as through only valves with roller or plunger actuating controls were used as limit switches. Additionally, it must be ensured that signals which influence one and the same cylinder are drawn beneath each other.

Motion and control diagram

Fig. 2/81

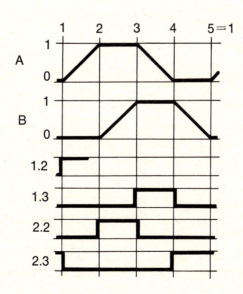

95

Fig. 2.81 shows that for this circuit no signal cut-out is required, provided that signal element 1.2 no longer emits a signal shortly before step 3. Since this signal element is a manual control, however, this condition is not necessarily certain.

When checking, it must be ensured that no signals influencing the same cylinder are present at the same time, as this would lead to a condition which is not uniquely defined, i. e. one signal blocks the other.

In this example, this could occur only with signals 1.2 and 1.3. It is then necessary to check which signal is required at which time and whether signals which are present with the system in its starting position, such as 2.3 in this case, have an effect on the control. Since 2.3 initiates the return movement of cylinder B, but cylinder B remains in the starting position and remains in the rear end position until 2.3 is freed, 2.3 does not cause a disturbance. The circuit is shown in Fig. 2/82.

Fig. 2/82

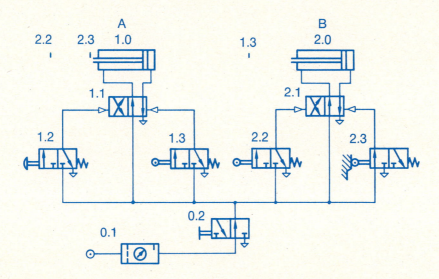

In Fig. 2/83, an interlock for 1.2 is incorporated by means of an additional limit switch at the rear end position of cylinder B. This ensures that the circuit remains operable over the whole working cycle when 1.2 is operated.

Fig. 2/83

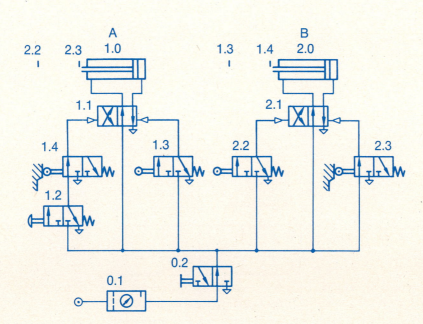

2.6.2 Exercise 6.2 Riveter

Problem:

Two clips are to be riveted together on a semi-automatic press.
Components and rivet are positioned by hand and then removed by hand on completion of the riveting operation. The automated part of the working cycle consists of the holding and clamping of the components (cylinder A) and also the riveting (cylinder B), and the cycle should be performed ending at the starting position after operating a start button.

Positional sketch and determination of working elements:

Fig. 2/84

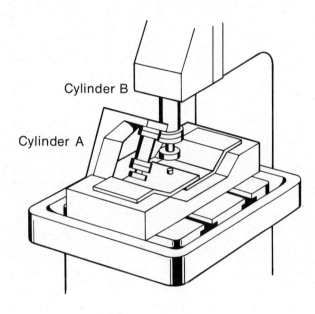

Displacement-step diagram:

Fig. 2/85

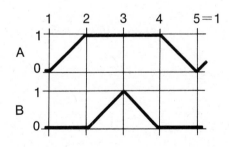

2.6.2.1 Construction of circuit diagram for exercise 6.2 with signal cut-out by means of idle return rollers

In constructing the circuit diagram, it is advisable to observe the points listed under 2.6.1.1 and to maintain the order given during this procedure.

The check to determine which signals are to be cut out can be made with the aid of the motion and control diagram (Fig. 2/86).

Motion and control diagram:

Fig. 2/86

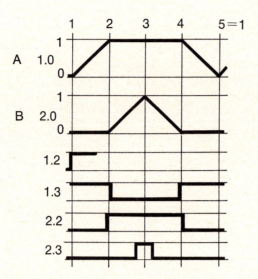

This shows clear overlapping of signals 1.2 and 1.3 and also of signals 2.2 and 2.3.
Owing to the first overlap, the system cannot start. The second overlap blocks the return stroke of cylinder 2.0. Thus, 1.3 and 2.2 must be cut out. Fig. 2/87 shows the circuit with signal cut-out by means of idle return rollers.

Fig. 2/87

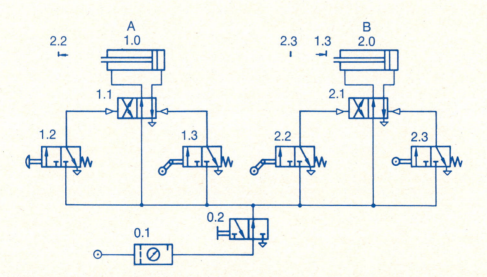

Since the start signal is not interlocked in this circuit, the control can be interfered with during operation by the start button.

It is recommended to provide the start interlock function by using the signal from the limit switch at the rear end position of the cylinder that performs the last movement in the sequence of movements. The interlock function is obtained by connecting the limit switch and the start switch in series.

The check to determine whether this interlock function is effective over the whole range can here again be made on the basis of the motion and control diagram (Fig. 2/88).

Fig. 2/88

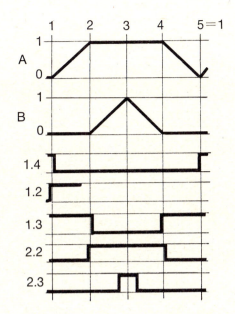

Fig. 2/89 shows the circuit with the fully interlocked start switch.

Fig. 2/89

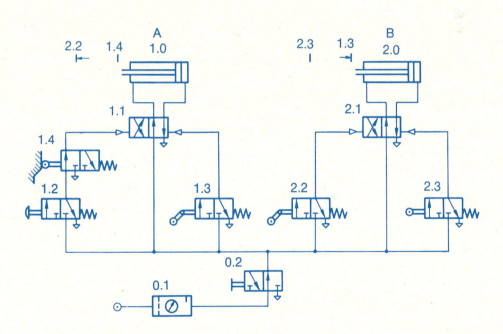

2.6.2.2 Construction of circuit diagram for exercise 6.2 with signal cut-out by means of reversing valve

If idle return rollers are not to be used to provide signal cut-out, an additional reversing valve is introduced. In Fig. 2/90, this is valve 0.2. Difficulties are presented in this case above all by the need to reverse 0.2 at the proper time.

It can be seen from the motion and control diagram (Fig. 2/88) that the signals from 1.3 and 2.2 must be cut out no later than the time at which the opposing signals for the respective cylinder are received. For this reason, it will be necessary to cut in and cut out respectively these signals 1.3 and 2.2 at steps 3 and 5; in the region up to step 3, signal 2.2 is required, and between step 3 and step 5 signal 1.3 is required. For reversal, signal 2.3 exists at step 3 and signals 1.2 and 1.4 exist at step 5. Limit switch 1.4 is incorporated here to latch the start button.

If a signal, or the energy supply itself, is to be distributed from one point it is advisable to draw distribution lines in the drawing (see outlets from valve 0.2 in Fig. 2/90). This clarifies the circuit diagram considerably.

Fig. 2/90

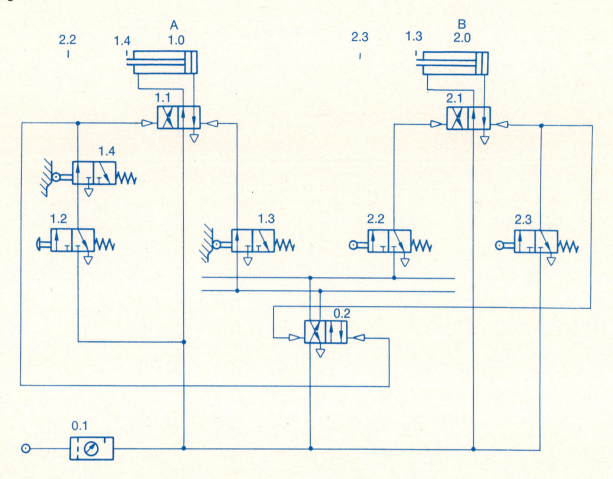

The circuit shown has the disadvantage, however, that two different operations can be triggered from one signal element (2.3 and 1.4). This could possibly lead to faulty switching.

This can be avoided by the signals from 1.4 and 2.3 not initiating both operations simultaneously but first of all reversing the reversing valve 0.2, and subsequently the outputs of this valve cause triggering of 1.1 and 1.2, see Fig. 2/91.

Fig. 2/91

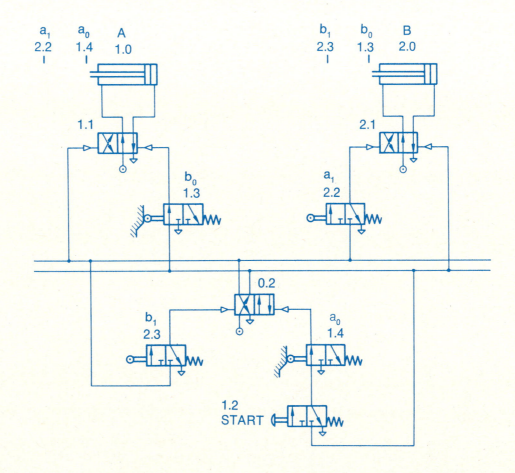

This example shows quite clearly however the limits of intuitive circuit diagram design. If the signals are to be cut off by means of reversing valves, an elaborate control can be designed only with a great deal of experience.

It is therefore advisable to apply an approach which permits a purely methodical procedure and which, above all, always results in a well-functioning control.

The simplest means of obtaining a dependable control is by simply cutting out each signal when it is no longer required. For the present example, this would mean cutting out one signal at each step, i.e. a total of 4 signal cut-outs.

Fig. 2/92 shows in block form such a unit with 4 signal cut-outs.

Fig. 2/92

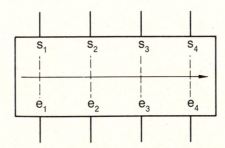

e_1 to e_4 represent the input signals, s_1 to s_4 the output signals.

Naturally, certain conditions must be satisfied if this unit is to solve the problem of the applied signals.

– Number of input signals = number of output signals.

– One output signal must be assigned to each input signal.

– The output signals must be stored, i.e. the desired output signal must be issued even when the corresponding input signal no longer exists.

- Only one output signal may exist at any one time, or it must be possible to eliminate specific output signals.
- The input signals may only ever be effective in the same sequence, namely 1 – 2 – 3 – 4 – 1 –...

If such a unit is available, the circuit can be made up as shown in Fig. 2/93:

Fig. 2/93

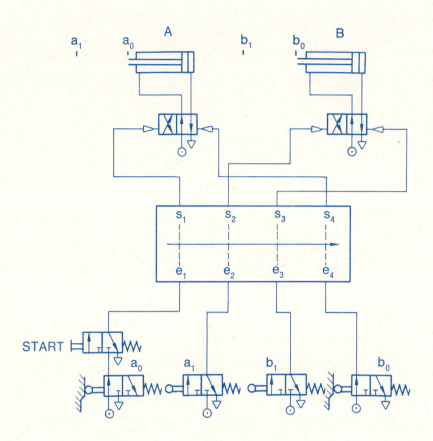

© by FESTO DIDACTIC

In designing the control, it is better to designate the limit switches by letters, namely a_0 and b_0 for the limit switches in the rear end position of the cylinder and a_1, b_1 for the limit switches in the forward end position.

Here again, the start signal is latched by the limit switch in the rear end position of cylinder A, i. e. by means of a_0.

The next problem is to find a circuit for the "block" which is used and satisfying all the conditions which are imposed.

Two of the possible designs are dealt with below:

– cascade
– shift register

2.6.2.3 Design of a cascade

Fig. 2/94 shows a circuit incorporating 4/2-way valves and satisfying almost all conditions imposed on the block with respect to signal cut-out. The designation "cascade" has arisen from the stage-like series connection.

Fig. 2/94

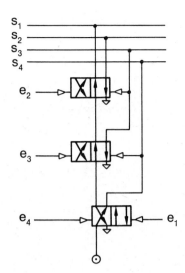

This arrangement ensures that at any one time only one output line is pressurized and all others are vented to atmosphere.

A further characteristic is the clear relationship between the inlets e and the outlets s, and also the sequence 1 . . . . . . n of the switching across the cascade.

With this arrangement, the signal cut-out can now be incorporated in a control with relative ease.

It must however also be ensured that an input signal which exists over a lengthy period of time cannot cause confusion in the circuit. This can be achieved by arranging for the input signal e_n to pass on only if an output signal s_{n-1} has previously been applied.

This can be accomplished in the circuit if, for example, a two-pressure valve is connected to the input to which the signals e_n and s_{n-1} are applied.

Fig. 2/95 shows the possible arrangements with two-pressure valves.

Fig. 2/95

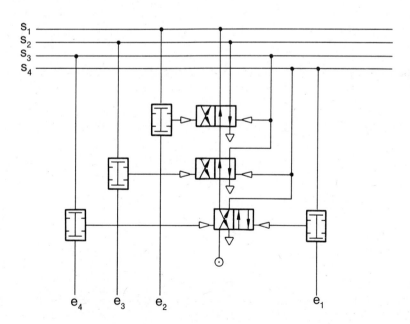

If the lines to the particular signal element – e. g. limit switch – are not too long, and if this signal element is not required to initiate another operation, then the two-pressure valves can be omitted and replaced by connecting in series the output s_{n-1} and the input signal e_n which triggers this output, as shown in Fig. 2/96.

Fig. 2/96

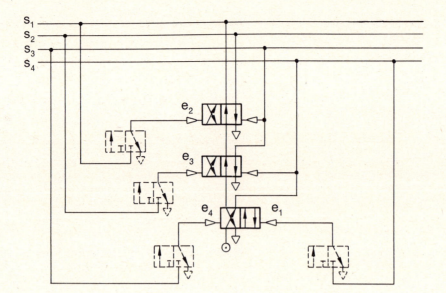

The various stations when switching through the cascade are shown in Fig. 2/97.

Fig. 2/97

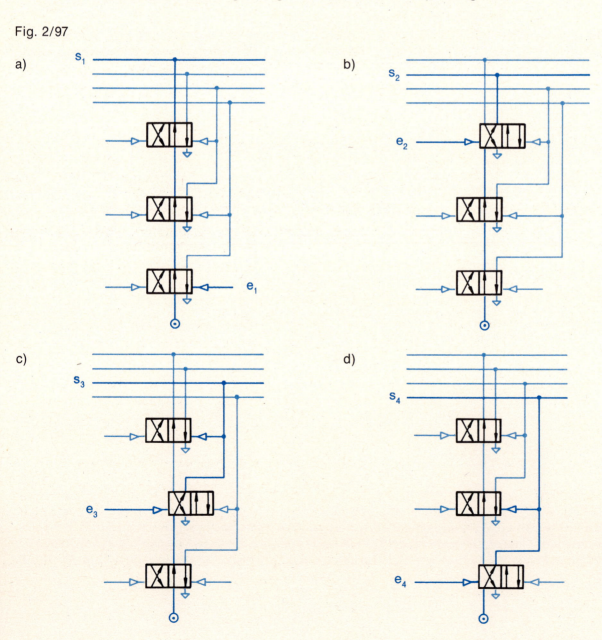

In principle, this cascade can be expanded to any number of stages. The arrangement always remains the same – all valves are connected in series, the 1st valve in the series issues 2 output signals s_1 and s_2 and all other valves issue one output signal each. The next valve in the series resets the valve before it, and so on. The last valve in the series receives two input signals so that there is always a uniform starting position – drawn in the control position which is opposite to the other valves.

Fig. 2/98 shows, as an example, a 3-stage cascade.

Fig. 2/98

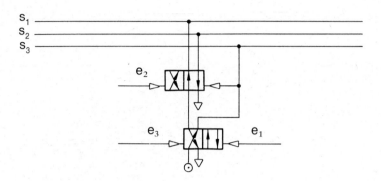

One point should however still be made: the limits of the cascade circuit are given by the characteristic that the energy is input through a single connection. Thus, the air must pass through all valves in the cascade before a control operation is initiated. The pressure drop resulting from this will be more noticeable if a larger number of valves are connected in series and the control becomes slower. In such cases it is advisable to make use of a shift register control.

2.6.2.4 Design of a shift register

In contrast to the cascade, 3/2-way valves are used which are not connected in series; here, each valve is connected directly to the air supply. Thus, the above-mentioned pressure drop does not occur when there are a large number of stages, but one valve more is always required compared with the cascade.

Fig. 2/99 shows a possible design for 4 stages. Other designs of shift registers are given in the volume "Control Engineering 2".

So that only one output signal can be issued, each stage is reset by the stage which follows it.

Fig. 2/99

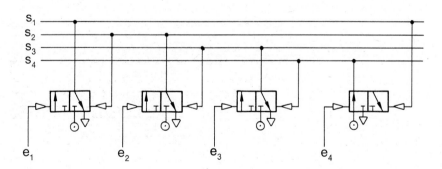

Here too, in order to latch the input signals, either a two-pressure valve must be connected ahead of the input and signals e_n and s_{n-1} must be applied to this valve or, if possible, output s_{n-1} and the signal element for input e_n must be connected in series. Figs. 2/100 and 2/101 show these possibilities.

Fig. 2/100

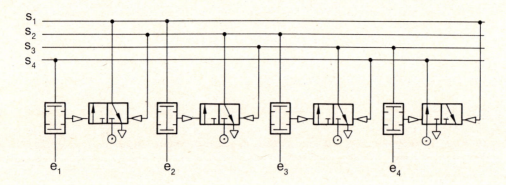

Fig. 2/101

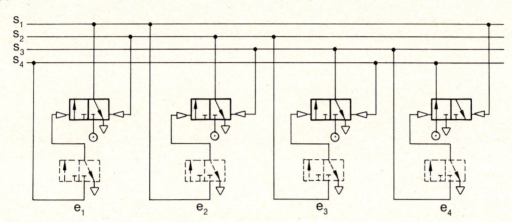

The possible control positions for this 4-stage arrangement are shown in Fig. 2/102.

Fig. 2/102
a)

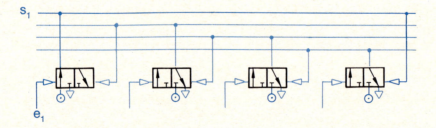

b)

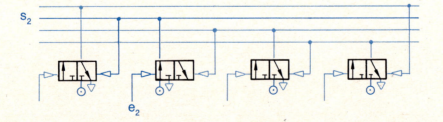

c)

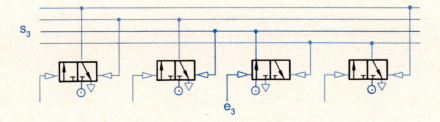

© by FESTO DIDACTIC

d)

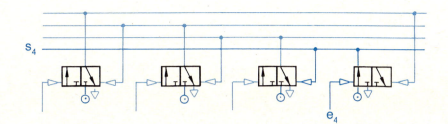

This arrangement can be expanded to any size. The important thing is that in the starting position the last valve is fitted in the operated position so that the 1st stage is primed to pass on the signal.

2.6.2.5 Method for designing coordinated motion controls with cascade or shift register

If a circuit is available for the cascade or shift register, it is an easy matter to design a coordinated motion control if one stage is used for each step, i. e. if each signal is cut out after the next operation has been triggered, regardless of whether this signal cut-out is necessary or not.

Fig. 2/103 shows a circuit for exercise 6.2 with signal cut-out being effected through a cascade.

Fig. 2/103

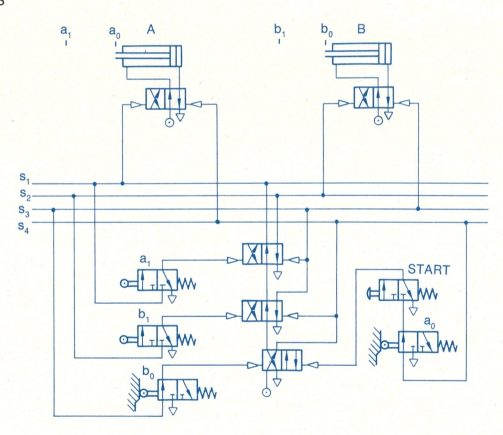

Fig. 2/104 shows the same circuit with signal cut-out through a shift register.

Fig. 2/104

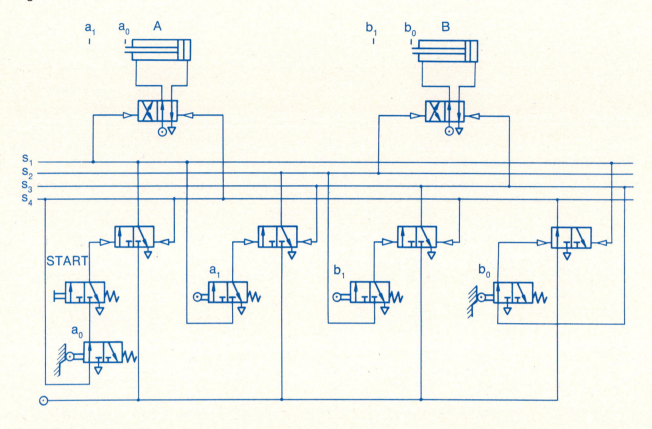

Owing to the basic design, this version is easy to deal with and above all each step is latched, i. e. only that operation can be triggered which is the next in line in the sequence. The sequencing of this control is dependable.

This type does however require a maximum of equipment. Frequently, partial latching of the installation is sufficient. In this case, the amount of equipment required for the circuit can be considerably reduced by cutting out only those signals which actually overlap.

The signals to be cut out can be determined either in the control diagram or, simpler still, by dividing them into groups in the motion sequence. No advance and return signals for one and the same cylinder may appear here within one group. For exercise 6.2, 2 groups can be formed:

A + B + / B − A − /
Group 1 / Group 2 /

This means for the controls that two reversals are necessary. Between this minimum configuration and the maximum configuration (one group per step), the circuit can be modified as required to suit the requirements.

In smaller controls, or with a smaller number of groups, as a rule the cascade is used and where there is a larger number of groups the shift register is used.

In the sections below, a methodical procedure for devising the circuit will be shown, based on the minimum configuration for the control and on the cascade as the basic circuit for the reversal block. The procedure given here can of course be applied similarly to expanded circuits and to the shift register as the circuit for the signal cut-out block.

2.6.2.6 Procedure for composing a circuit diagram by the Block Method

1. Definition of the sequence of movements in the motion diagram and in the abbreviated notation. For exercise 6.2, the motion diagram has already been established (see Fig. 2/85). The abbreviated notation is

   A + B + B − A −

2. Division into groups:
   To end up with the minimum number of reversing valves, the sequence is divided into groups. In this division, it must be ensured that each cylinder occurs only once in each group. 2 groups result for exercise 6.2

   A + B + | B − A − |
   1   2

   Feedback is necessary from the group formation.

3. Drawing of cylinders and associated impulse valves. When constructing the circuit diagram by the Block Method, the cylinders are generally triggered by impulse valves.

4. Designation of the elements:
   It is advisable to use the alphabetic type of designation for the draft.

5. Drawing of the cascade or shift register and allocation of outputs to inputs.
   The number of valves required is equal to the number of outputs required (see "Division into groups") minus 1.

   For exercise 6.2: 1 valve (Fig. 2/105)

   Designation of outputs: $s_1, s_2 \ldots s_n$
   Designation of inputs: $e_1, e_2 \ldots s_n$

Fig. 2/105

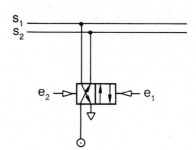

6. Allocation of outputs to the created control circuits.
   One should consider whether a control circuit change takes place in the starting position.
   If switchover is required, as in exercise 6.2, the last group is pressurized in the starting position of the system.

   A +, B + | B −, A − |
   s_1 s_2

7. Transposition of the motion diagram into the circuit.
   When the preparatory steps 1–6 have been completed, the draft design may be commenced. The following must be noted:

7.1 The motion diagram is transferred step by step.

7.2 First of all, it must be checked whether the cascade or shift register need be reversed at the step concerned:

   – if yes: the applied signal is connected directly to the input of the block (switchover to the next output). The output signal which now exists is used directly for initiating the next movement which is to be performed. The air for the signal element is also taken from the outputs, namely from the control circuit which is switched off.

- if no: the signal element operated at that step initiates directly the next action and is pressurized by the output of that control circuit which is currently conveying air.

7.3 If a cylinder is to travel out several times during a motion cycle, the limit switches assigned to the cylinder are also operated several times, and in the course of this another operation is to be initiated with each actuation. This means that these limit switch signals must be interlocked. This is effected by means of a two-pressure valve which is pressurized by the line pressurized at the time and by the signal element.

7.4 Auxiliary conditions and additional interlocks are incorporated only when the basic sequence of motions has been completely designed. It is definitely advisable to proceed step by step.

Items 7.3 and 7.4 are not required for exercise 6.2, but they have been included at this point for the sake of completeness.

Fig. 2/106 shows the circuit for exercise 6.2 drawn up by the Block Method. Here too, the start signal is locked by means of the additional limit switch a_0. If the circuit diagram has already been drafted, the digital type of designation may be used for the final version. (Depending on the extent, either consecutive numbering or classified digital designations as in Fig. 2/106.)

This circuit, incidentally, corresponds to the intuitively designed circuit shown in Fig. 2/90: thus, the method automatically leads to the control with the simplest function.

Fig. 2/106

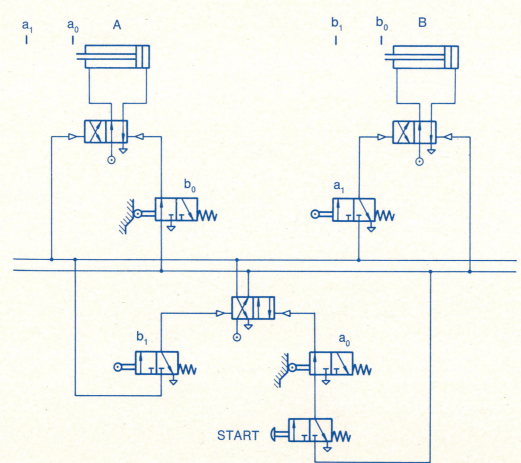

2.6.3 Exercise 6.3 Stamping appliance

Problem:

Rectangular parts are stamped on a special machine. The parts are taken from a gravity-feed magazine, pushed into the machine against a stop and clamped by means of a cylinder, stamped by a second cylinder, and ejected by an ejector cylinder.

Positional sketch and arrangement of the cylinders:

Fig. 2/107

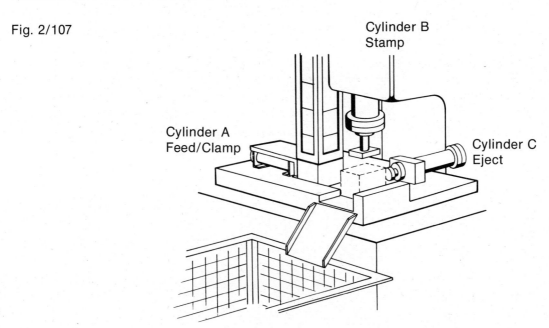

Displacement-step diagram:

Fig. 2/108

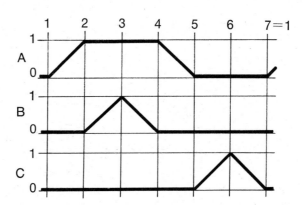

Auxiliary conditions:

1. The sequence of operations should be performed automatically, the choice being provided for:
   - single cycle
   - continuous cycling

   The start signal is input by means of a start button.

2. The magazine should be monitored by means of a limit switch. If there is no part left in the magazine, the system must be stopped in its starting position and interlocked to prevent restarting.

3. If an EMERGENCY STOP switch is operated, all piston rods of the three cylinders should return immediately from any position to their starting positions and be operable again only when the interlock has been removed.

2.6.3.1 Construction of circuit diagram for exercise 6.3 with signal switch-off via idle return rollers

First of all, the circuit diagram is drawn up by following the known procedure and without including any additional requirements, i. e. for one working cycle.

Fig. 2/109 shows the motion and control diagram. One can see from this that signals 1.3, 2.2, and 3.2 may not exist at all times, that is they must be deleted some of the time, and that disturbances can enter the system through 1.2.

Fig. 2/109

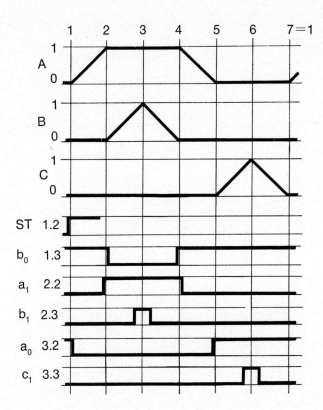

© by FESTO DIDACTIC

Fig. 2/110 shows a circuit with signal switch-off via idle return rollers, but without interlocking of start button 1.2 which in this sequence of motions can no longer be completely locked by the limit switch.

As already stated, the incorporation of the additional requirements should be accomplished step by step. In the circuit shown in Fig. 2/111, valves 1.2, 1.6, and 1.8 are necessary to satisfy requirement 1. Requirement 2 is satisfied by valve 1.10; here, the whole system remains stationary in its starting position when the magazine is empty, and the start signal is blocked.

With regard to the EMERGENCY STOP requirement, it is necessary to pressurize the main valves on the "forward travel side" and to ensure that there is no longer a signal on the opposite side. This is achieved by removing via valve 0.3 the air from the valves controlling the forward travel side if the EMERGENCY STOP is operated.

Fig. 2/110

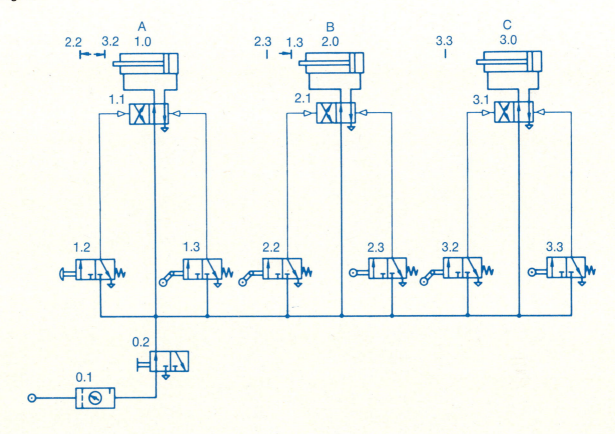

Fig. 2/111

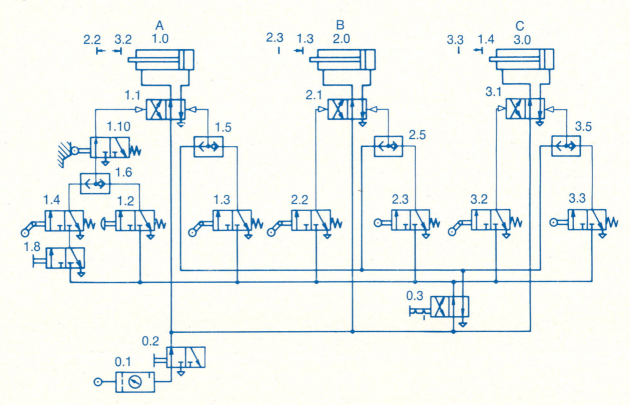

113

2.6.3.2 Circuit diagram for exercise 6.3 with signal switching via reversing valve. Design by the Block Method.

With a control of this size, it is best to use the cascade method for signal cut-out. The procedure is demonstrated below.

Sequence of motions in abbreviated notation and division into groups.

A +, B +, / B −, A −, C +, / C −
1   2   1

Two groups result for this circuit (cyclic group formation). This means that a 4/2-way valve is required for the cascade. The complete circuit is shown in Fig. 2/112. Since with this group division it is not necessary to reverse the cascade in the starting position of the system, the Start button is placed between line 1 and valve 1.1. To obtain complete interlocking of the start signal over the entire cycle, limit switch 1.4 (c_0) must be built in additionally for interlocking via the control circuits since the cascade switches off the start signal only from step 3 to step 6.

Fig. 2/112

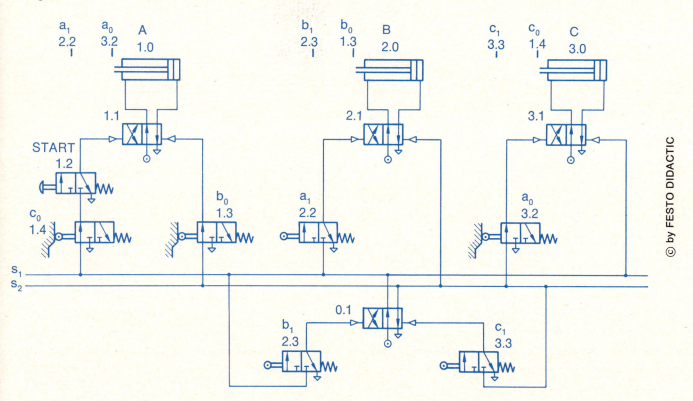

This circuit does however have the disadvantage that when operating the limit switches a_1 and b_1 in the starting position, the control can be switched incorrectly. The circuit under consideration is thus not completely blocked in its starting position against outside influence. This is because the cascade in its starting position already activates the signal elements through which the next commands are output.

If this is to be prevented, the circuit must be extended. By including an additional control circuit, one can obtain interlock in the starting position of the system. An important point is that the cascade is switched over only by the start signal and consequently that the control circuit is pressurized in the starting position which influences the final motions in the sequence of operations.

In the present example, we then have the following groups:

A +, B +, / B −, A −, C +, / C − /
1   2   3

Thus, three reversals are required which means that two impulse valves must be used. Fig. 2/113 shows this circuit.

In this circuit, a complete interlock exists only in the starting position. If this interlock is to be expanded to cover the whole sequence, the cascade must be extended such that one control circuit is available for each step in the sequence of motion. This ensures that at any one time only the signal element which must output the next command is pressurized.

Fig. 2/113

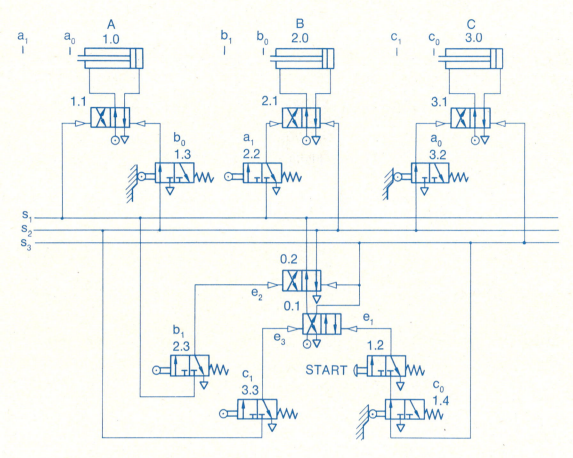

Incorporation of the additional requirements (auxiliary conditions) is accomplished step by step in the manner described above.

Fig. 2/114 shows the circuit with the incorporated auxiliary conditions (as in circuit shown in Fig. 2/111). It should however be noted at this point that the system starts immediately when switching over from single cycle to continuous cycling, i. e. the selector switch is also the start switch for continuous operation.

This disadvantage is typical for the type of circuit under consideration if a valve with fixed roller is used as signal emitter for automatic operation.

Fig. 2/114

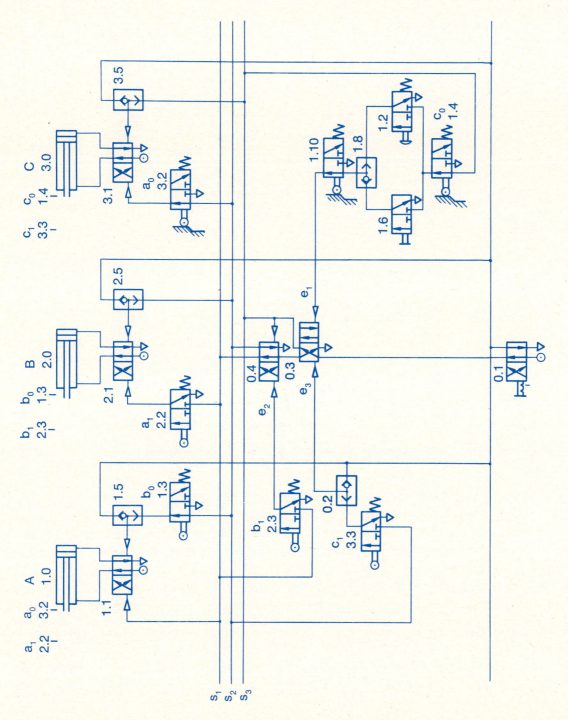

Furthermore, the circuit under consideration does not necessarily comply with present-day requirements in terms of convenience of operation. Since increasing use is being made of control panel and control cabinet installations, it must be possible to fit the operator control elements and operate them easily.

Fig. 2/115

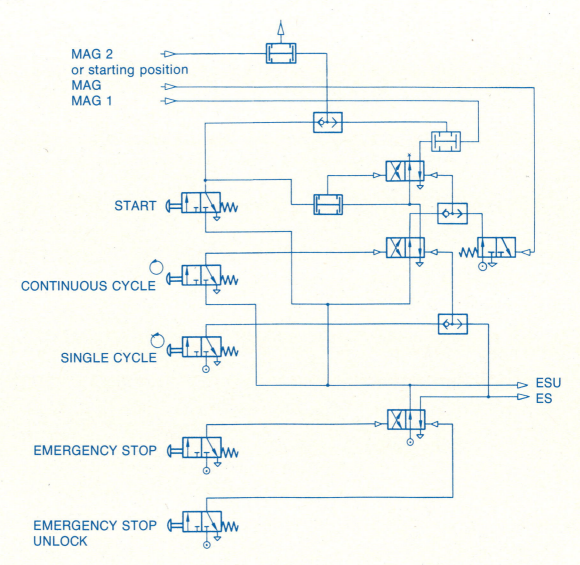

Fig. 2/115 shows a possible circuit for the imposed conditions. In this circuit, push buttons are used as operator control elements.

The operating mode is selected by the "Continuous Cycle" or "Single Cycle" push buttons. The motion sequence commences after the START button has been operated, either ○ or ○.

Through the MAG, MAG 1, MAG 2 connections, the circuit can be locked in the initial position.

If MAG 1 is connected, latching is provided only for the "Continuous Cycle" position; "Single Cycle" can be started even when the magazine is empty.

If MAG 2 is connected, the entire control is latched, i. e. if the magazine is empty it is not possible to start.

The output must be latched with the "Starting Position" signal, i. e. limit switch in the rear end position of the cylinder that has performed the last movement in the sequence of movements, and possibly with the relevant line in the cascade or shift register.

The input circuit is put out of operation by the Emergency Stop button, and at the same time the EMERGENCY STOP and EMERGENCY STOP UNLOCK signals can be picked off from the NS and NSE connections. If the circuit is in the "Continuous Cycle" mode, then the emergency stop signal switches it over to single cycle.

If valves with a locking position are used as input elements instead of push buttons, the control is simplified.

As an example. Fig. 2/116 shows an expanded input circuit with selector switches.

Fig. 2/116

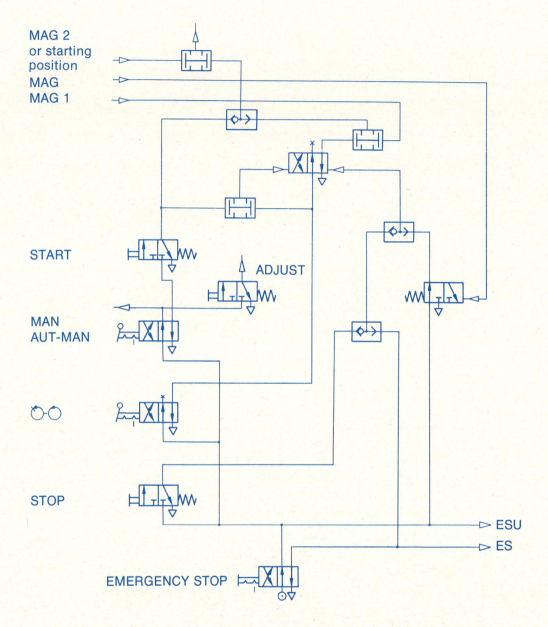

In this circuit, it is also possible to select AUTOMATIC and MANUAL operation. In the MAN mode, each cylinder can be controlled individually, for example by means of additional push buttons. The circuit for AUT operation is then ineffective. By means of the "Adjust" push button, it is possible, for example, to influence a specific shift register in the MAN mode.

The other conditions are the same as for the circuit shown in Fig. 2/115. These examples for input circuits should illustrate that partial problems can be extracted from total circuits and these can be designed as unit circuits. Fig. 2/117 shows how this unit circuit is incorporated for the input conditions in example 6.3.

Fig. 2/117

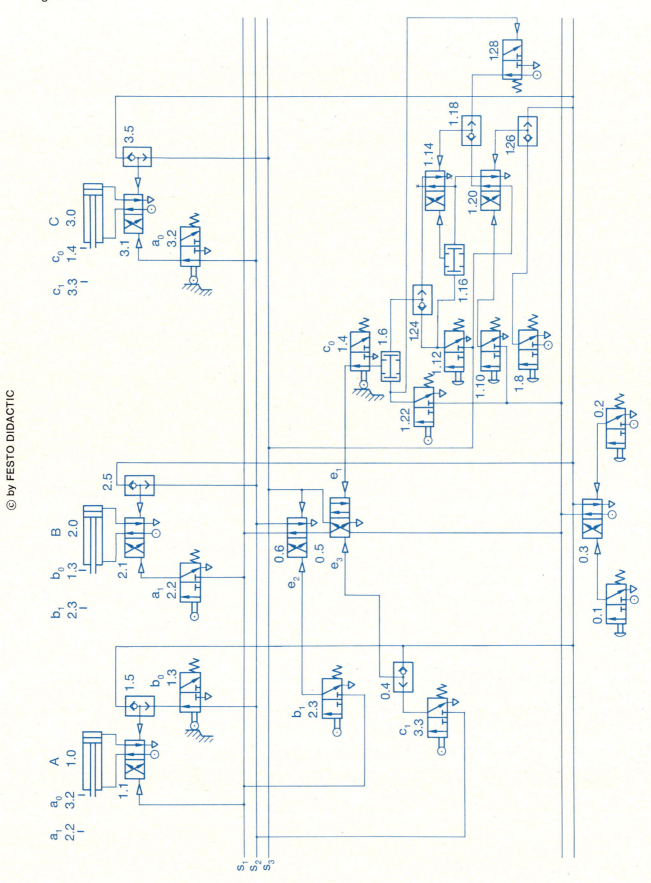

2.6.4 Exercise 6.4 Bending fixture

Problem:

Metal sheets are to be flanged on a pneumatically operated bending tool. After clamping the component by means of a single-acting clamping cylinder A, it is bent over by a double-acting cylinder B and subsequently finish bent by another double-acting cylinder C. The operation is initiated by a manual button. The circuit is to be designed such that one working cycle is completed each time a start signal is given.

Positional sketch:

Fig. 2/118

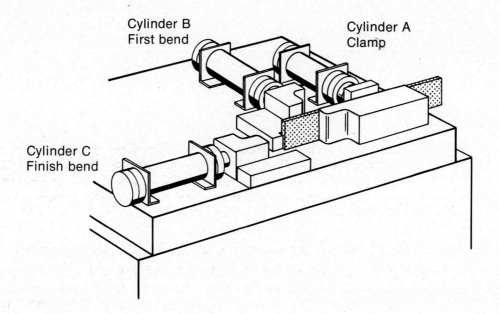

© by FESTO DIDACTIC

Displacement-step diagram:

Fig. 2/119

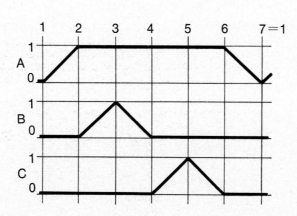

2.6.4.1 Circuit for exercise 6.4 with signal cut-out via idle return rollers

In order to examine the signals to be switched off, the control diagram is drawn up (Fig. 2/120).

Fig. 2/120

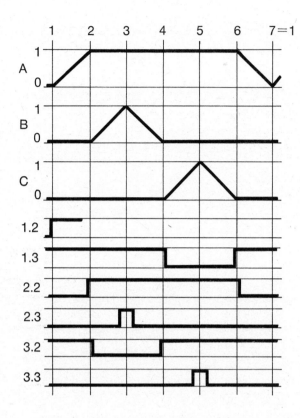

It can be seen that provision must be made for switching off signals 1.3, 2.2, and 3.2. In the circuit shown in Fig. 2/122, an interlock is incorporated for the start signal, and this is provided by the additional limit switch 1.4 at the rear end position of cylinder 1.0.

It must be checked individually whether this provision actually ensures complete interlocking of the start signal.

Fig. 2/121

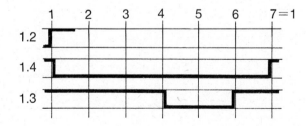

It can be seen from Fig. 2/121 that 1.4 is present only with steps 1 and 7, i. e. at the time at which the start signal is required.

Fig. 2/122

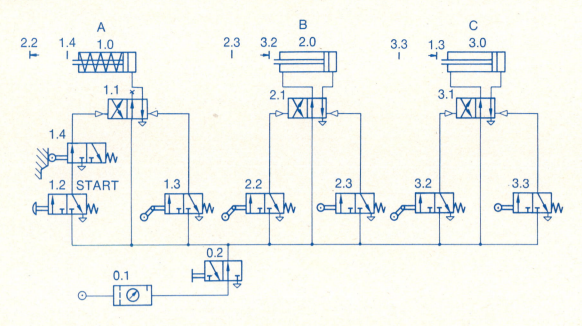

2.6.4.2 Circuit for exercise 6.4 based on the Cascade Method

Sequence of motions in abbreviated notation and division into groups:

A +, B + | B −, C + | C −, A − |
1 | 2 | 3 |

In constructing the circuit diagram, the sequence described under 2.6.2.3 is adopted. Fig. 2/123 shows the resulting circuit.

Here too, start signal interlocking is effected by means of a limit switch a_0 and the control circuit. In the sequence of motions under consideration, interlock by means of the limit switch would however be sufficient and hence it would be possible to do without the additional facility provided by the control circuit and the air for the start button could also be taken directly from the network.

Fig. 2/123

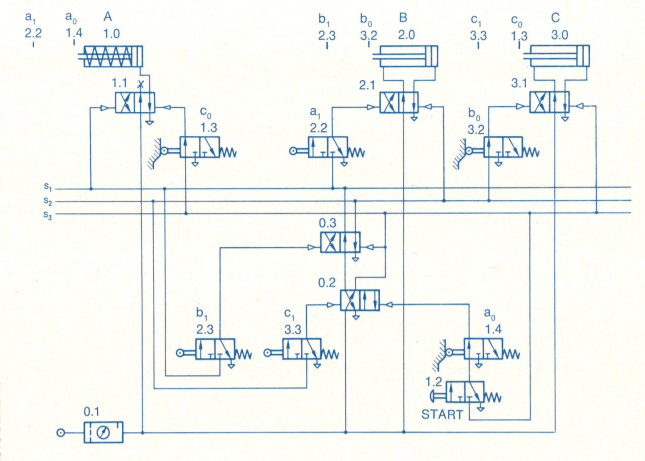

2.6.5 Exercise 6.5 Shearing unit

Problem:

Bar stock is cut to length on a shearing fixture. The feed is by means of a pneumatic cylinder B and at the same time this moves a pneumatic clamping cylinder A during the feed motion. When the material has been pushed against a fixed stop, it is securely held by a clamping cylinder C while the clamping gripper releases and the feed cylinder travels back. When the material has been sheared by means of cylinder D, the clamping cylinder is also released and a new sequence of operations can commence.

Auxiliary conditions:

1. The circuit must be designed for single cycle and continuous operation.
2. A new working cycle may be initiated only when all cylinders have reached the rear end position.

Positional sketch and arrangement of the cylinders:

Fig. 2/124

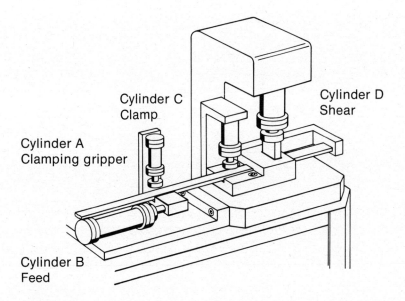

Displacement-step diagram:

Fig. 2/125

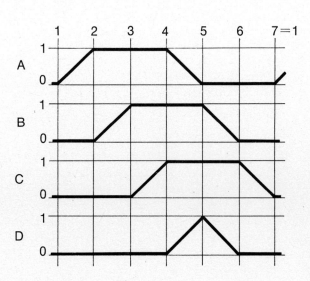

2.6.5.1 Circuit for exercise 6.5 based on the Cascade Method

Provided the clamping operations also can be verified by means of limit switches, it is possible to construct a circuit by the Cascade Method.

This circuit is shown in Fig. 2/126. Since in this example two operations are initiated simultaneously, this must be considered when dividing into groups.

$$\begin{array}{c|c|c|} A+, B+, C+ & A- & D-, C- \\ & D+ & B- \\ 1 & 2 & 3 \end{array}$$

Owing to the simultaneous initiation of two motions, the sequence is separated and must be rejoined at the end of the motion. Here, this is effected by means of limit switches b_0 and c_0.

Since in circuits of this type it is very difficult to uniquely assign the valve designations to the advance or return movement of a cylinder, consecutive numbering is introduced here for designating the elements. The designations of the cylinders and limit switches for the design remain unchanged.

Fig. 2/126

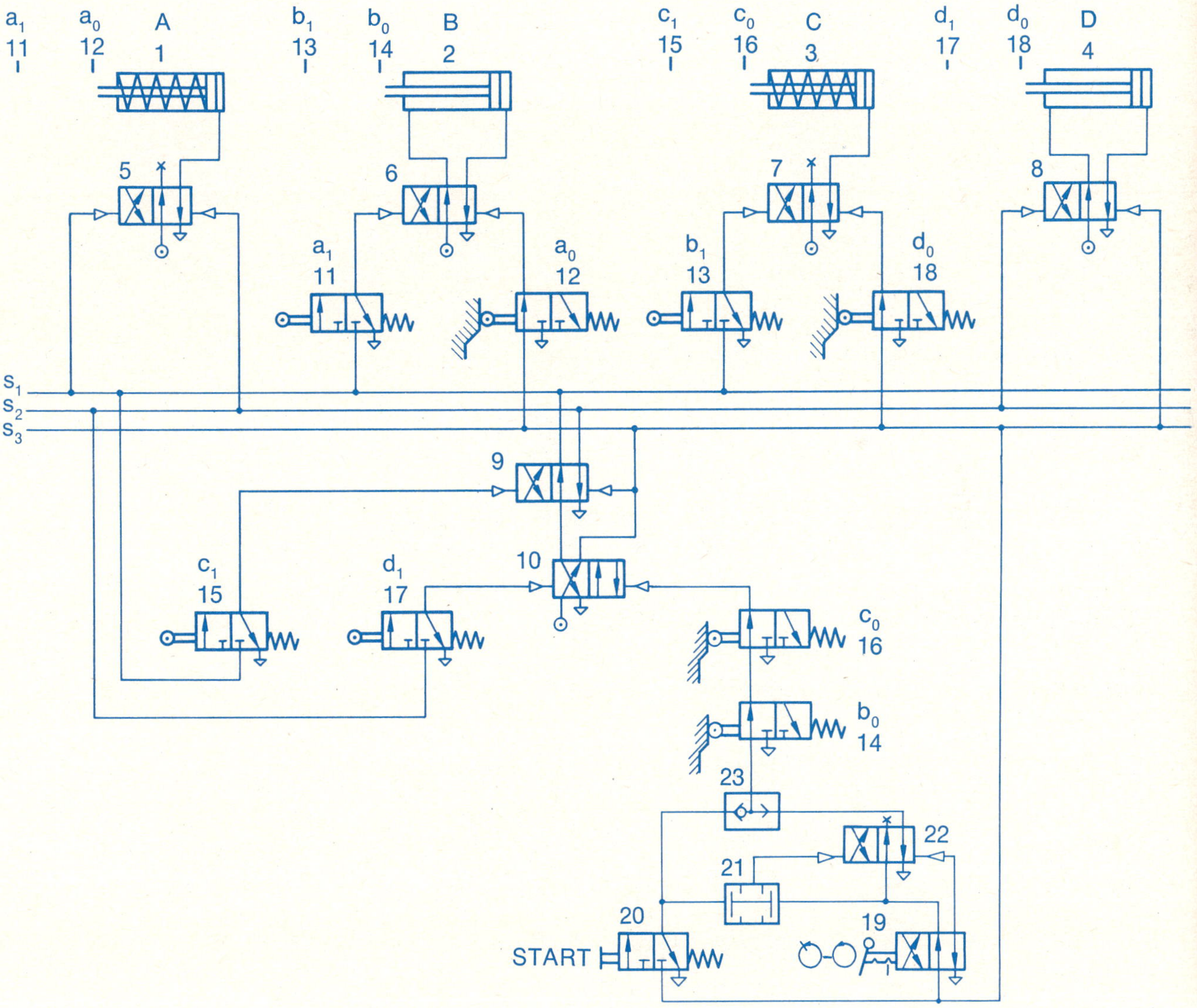

2.6.6 Exercise 6.6 Pressing fixture

Problem:

Parts are slowly pressed in on an asembly fixture. To attain the full depth of pressing, the press cylinder must operate a second time with a short impact and then hold the workpiece until a second cylinder has pressed in a locking pin from the side.

Positional sketch and arrangement of cylinders:

Fig. 2/127

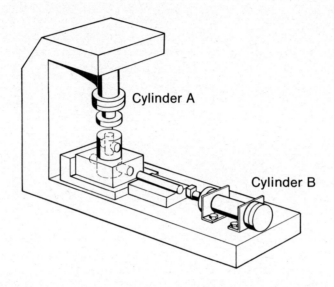

Displacement-step diagram:

Fig. 2/128

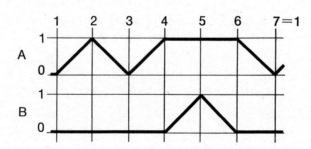

Displacement-time diagram:

Fig. 2/129

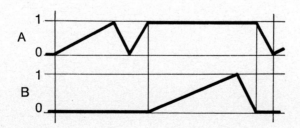

© by FESTO DIDACTIC

2.6.6.1 Circuit for exercise 6.6 with idle return rollers

Fig. 2/130 shows a possible solution with signal cut-out via idle return rollers. Since the piston of cylinder A travels out twice and the valve 1.3 mounted at the forward end position of this cylinder influences first the return motion of the piston of cylinder A and then the forward motion, signal reversal is necessary and this is effected by means of valve 1.9. Owing to the double return motion of cylinder A, valve 1.4 is also operated twice but it may not supply a signal the second time. This means that another reversal is necessary, and here valve 1.8 removes the air from 1.4. The reversing valve 1.8 is controlled by 1.2 and 1.5, and reversing valve 1.9 is controlled by 1.4 and 1.5.

To obtain the various forward travel speeds for cylinder A, valve 1.02 and one-way flow control valve 1.01 are incorporated. When the piston of cylinder A travels out the second time, 1.02 is reversed by 1.4 and the exhaust air can escape directly. i. e. it need not pass through the throttling point. This allows a higher speed to be reached. The return control of 1.02 is by means of a signal from 1.9.

Fig. 2/130

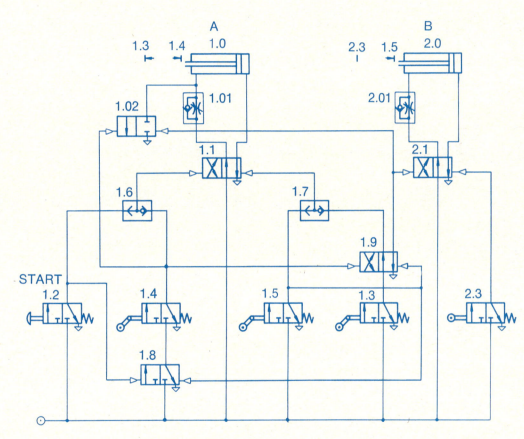

2.6.6.2 Circuit for exercise 6.6 with cascade and shift register

With this circuit too, it is possible to either latch fully, i. e. to incorporate a signal cut out for each step, or to cut out only those signals which in fact have to be cut out. Fig. 2/131 shows the fully latched circuit.

Fig. 2/131

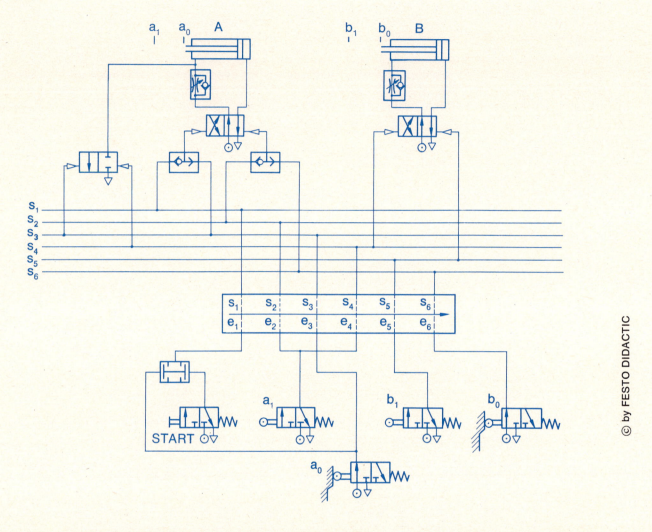

In this version, the block may be designed either in cascade or in shift register form. It is however necessary here to also include two-pressure valves for latching because the limit switches are connected directly to the network. Fig. 2/132 shows this circuit.

Fig. 2/132

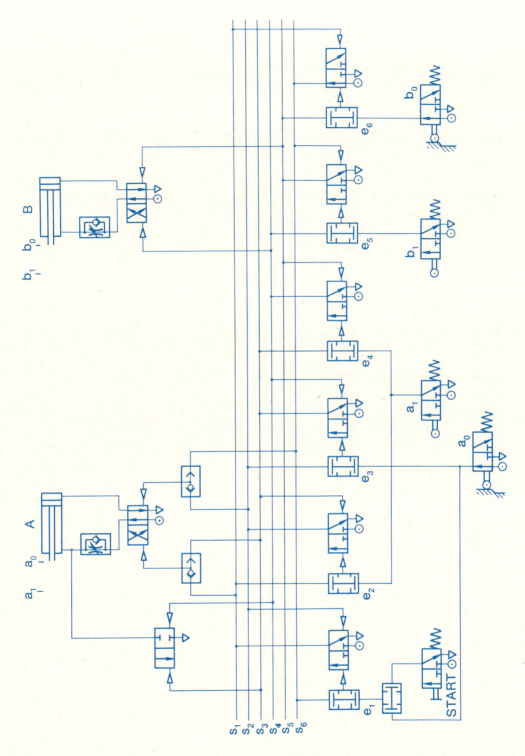

If the circuit is to be built with the minimum number of components, then it is necessary to divide into groups.

A +	A −	A + B +	B − A −
1	2	3	4

This circuit requires at least a four-stage signal cut out. Fig. 2/133 shows this version.

Fig. 2/133

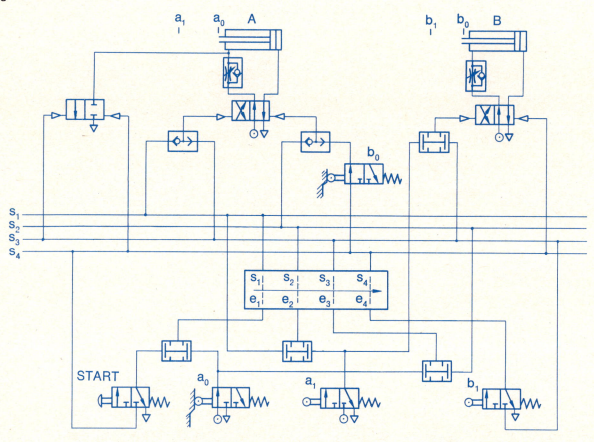

This circuit reveals another peculiarity of the simplified version. If a cylinder is triggered several times during a sequence of operations (as in the present case, cylinder A), it is necessary to provide additional interlocking of the multiply operated limit switches, in this case a_0 and a_1, and this can be effected by means of two-pressure valves (see item 7.3 in Section 2.6.2.3).

Here too, the block may be designed by the cascade or by the shift register method. Fig. 2/134 shows a design with the cascade.

Fig. 2/134

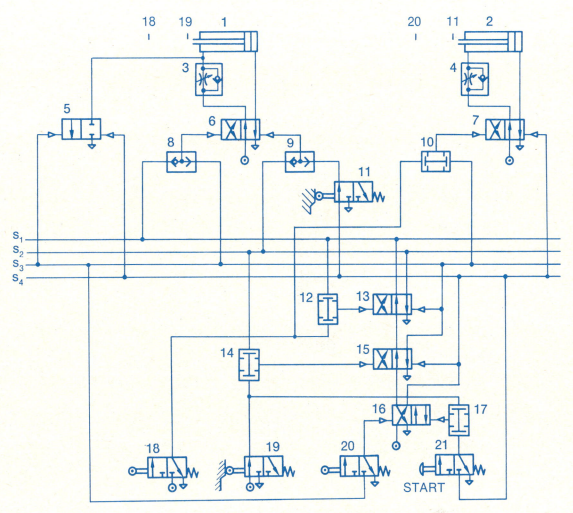

2.6.7 Exercise 6.7 Drilling unit

Problem:

A control is to be designed for a special machine on which two equally sized holes are to be drilled close together in rectangular parts.

The parts are taken from a gravity-feed magazine and pushed into the fixture against a fixed stop by a double-acting cylinder.
The parts are held securely by the feed cylinder during the machining operation. A pneumatic-hydraulic feed unit is used for the feed of the drilling spindle used in the machining operation.

A feed table operating between fixed limits by two fixed stops in conjunction with a double-acting cylinder is used for positioning the second hole.

The parts are ejected by an ejector mechanism which is operated on the return motion of the feed cylinder.

Auxiliary conditions:

1. When the "single cycle" selector switch is operated and after depressing the "START" button, the system should perform one working cycle and then remain stationary in the starting position.

2. On operating the "continuous cycling" selector switch, fully automatic operation of the system must be ensured after the start signal has been provided, until the opposing signal "single cycle" (stop in starting position) or requirement "3" is applied.

3. Magazine monitoring: if no more parts are in the magazine, the system should remain stationary in the starting position and at the same time block the start for "continuous cycling".

4. EMERGENCY STOP: on operating the "EMERGENCY STOP" button, all cylinders should return from any position to the starting position, but cylinder A and C only when cylinder B has reached the rear end position.

Positional sketch and arrangement of the cylinders:

Fig. 2/135

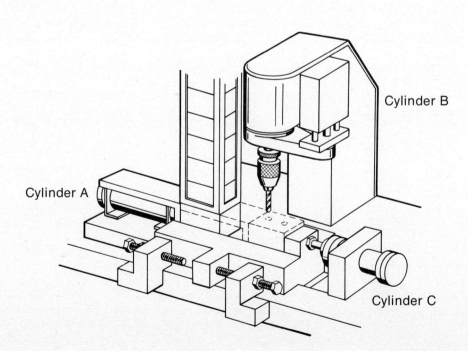

Displacement-step diagram:

Fig. 2/136

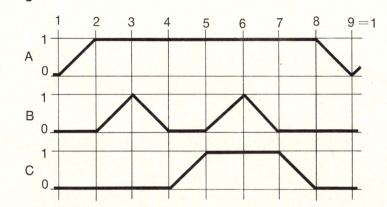

2.6.7.1 Solution to exercise 6.7

In this problem, it is advisable to use the block system approach for a circuit.

Signal cut-out by means of idle return rollers cannot be considered because in this problem overlaps can occur owing to the long overtravel distances and in addition a continuous signal from a limit switch must exist under all circumstances in the rear end position of cylinder B (auxiliarly condition 4).

Division into groups results in four groups or three reversing valves.

A +, B +	B −, C +	B +	B −, C −, A −
1	2	3	4

Fig. 2/137 shows a circuit without incorporated additional requirements.

In the circuit shown in Fig. 2/138, a modified standard circuit per Fig. 2/115 has been used to incorporate the additional requirements. The additional condition that cylinder 2 must first be in the rear end position if EMERGENCY STOP is applied is accomplished by means of an interlock via a two-pressure valve.

Fig. 2/137

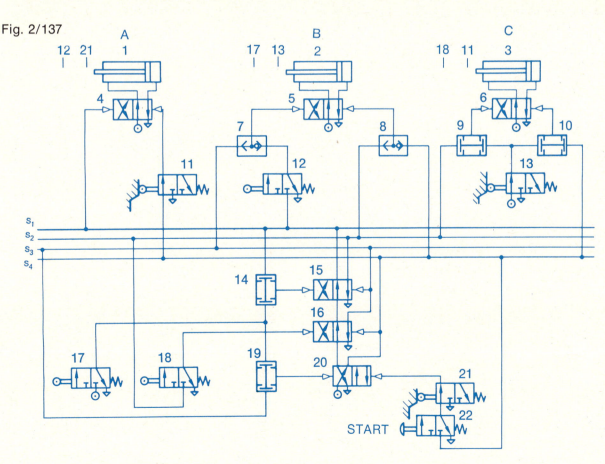

Fig. 2/138

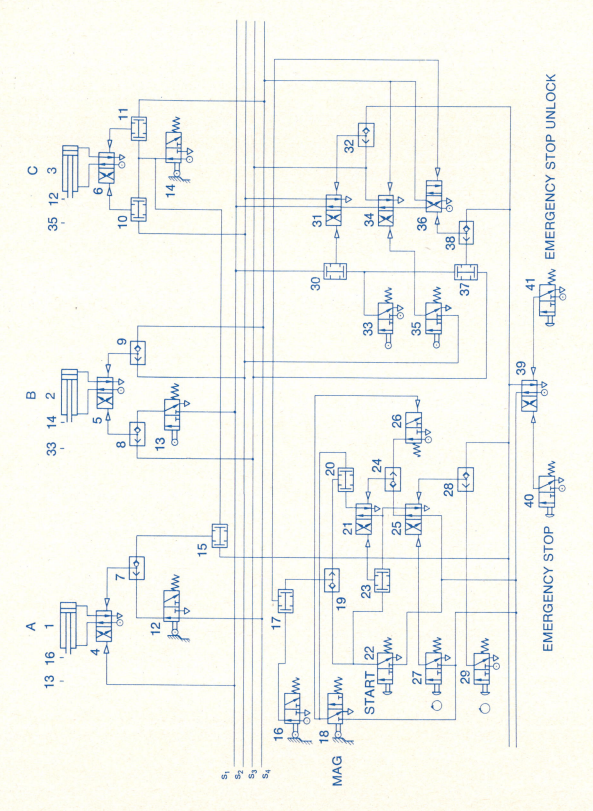

2.7 Constructing circuit diagrams for pilot controls

Pilot controls in pure form do not occur very frequently in practice. For the most part, they occur in combination with other types of control.

Since in a pilot control (sometimes referred to as a combinational control) the output signal always has a clear relationship to the input signals, it is important when designing to consider all possibilities with respect to the input signals. The most certain and quickest method of designing a circuit is one which adopts a logic approach (see the "Control Engineering 2" brochure).

If, as here, the circuit is to be designed intuitively from a knowledge of equipment, then the conditions for the outward movement of the cylinder should be noted in the form of a table or a listing. Each condition should then be transferred step by step into the circuit.

Exercises 7.1 and 7.2 give examples of pilot controls.

2.7.1 Exercise 7.1 Test station for cans

Problem:

Cans are to be checked at an inspection station to determine whether the lid is on. If a can is received without lid, the defective can is to be pushed to one side by means of a pneumatic cylinder. Lid and can are tested by means of reflex sensors.

Positional sketch:

Fig. 2/139

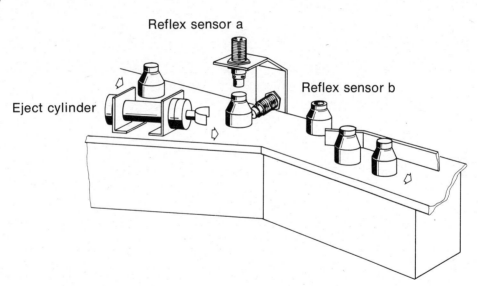

2.7.1.1 Circuit design for problem 7.1

The conditions given in the problem description can be put down in tabular form.
Reflex sensor a checks the lid. Reflex sensor b the can. The output signal z can be related to the various possibilities.

Lid	Can	a	b	Cylinder
Not present	Not present	0	0	Retracted
Present	Present	1	1	Retracted
Not present	Present	0	1	Travel out
Present	Not present	1	0	Cannot occur!

In simplified form, this table provides a list of the 4 possible combinations of two binary input signals in a control and the associated output signal.

	a	b	z
1.	0	0	0
2.	1	1	0
3.	0	1	1
4.	1	0	–

For the circuit, this means that the cylinder must be moved out by the signals a not operated **and** signal operated (see line 3.) Return control is effected in the circuit shown in Fig. 2/140 by a delayed reversal of valve 1.1 after the condition for outward movement has been removed, and this is obtained through the one-way flow control valve 1.7.

Fig. 2/140

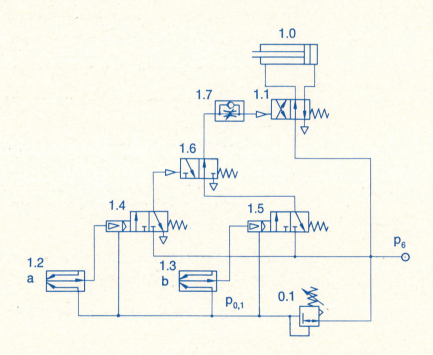

2.7.2 Exercise 7.2 Test station

Problem:

Components are to be checked for proper dimensions on an inspection device. Sensing is by means of 3 reflex sensors. 3 visual indicators show whether the component is correct, too small or too large.

		Reflex sensor	Visual indicator
Too small	(49.9):	a	z
Correct	(50.0):	b	y
Too large	(50.1)	c	x

Positional sketch:

Fig. 2/141

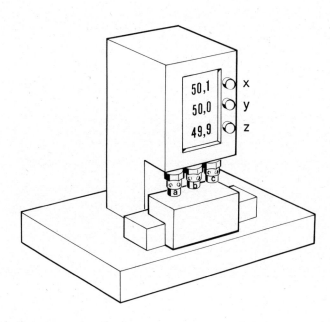

2.7.2.1 Circuit design for exercise 7.2

Since an indicator lights up only for the corresponding part, a logic circuit must be developed. Here too, the possible input combinations can be put together in a table.

Reflex sensor			Visual indicator		
a	b	c	x	y	z
0	0	0	0	0	0
1	0	0	0	0	1
1	1	0	0	1	0
1	1	1	1	0	0
0	0	1			
0	1	1	Combinations which in practice cannot occur and which thus cannot cause an indicator to light up.		
0	1	0			
1	0	1			

The circuit can be derived from this table. Fig. 2/142 shows a possible design.

The circuit in which a signal is issued if the reflex sensor does not respond can be obtained by means of a normally open 3/2-way valve connected in series on the downstream side.

The condition that a signal be issued only if several sigalling elements are operated can be obtained either by means of two-pressure valves or alternatively by connecting in series.

Fig. 2/142

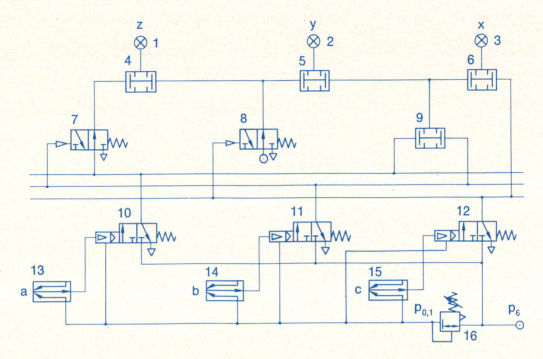

2.8 Constructing the circuit diagram for memory controls

In pure memory controls, one must pay particular attention to the storage and linkage of signals.

When constructing the circuit diagram, it is important to uniquely define the problem and the conditions and to use a suitable form of representation.

As it is not possible with memory controls to draw up motion diagrams (motion sequences are not produced automatically by the control but are input as desired), another form of representation must be selected. The conditions may be recorded in this case in tabular form.

It is not possible to specify a purely schematic circuit diagram design for memory controls; a solution must be sought by the intuitive method.

It is advisable to proceed roughly along the following lines:

– Problem breakdown

– Listing of conditions in tabular form

– Recording of the specified magnitudes in the circuit diagram

– Step-by-step incorporation of the individual requirements from the table which has been formed into the circuit diagram.

We would also point out that exercises 2.4.1.4, 2.4.1.5, and 2.4.1.17 are concerned with pure memory controls.

2.8.1 Exercise 8.1 Pneumatic switching section

Problem:

Workpieces arriving on a conveyor belt are to be distributed between four different belts by means of a pneumatically operated switching section. Shifting into the required position is initiated by means of four push buttons and must be possible in any sequence.

Positional sketch:

Fig. 2/143

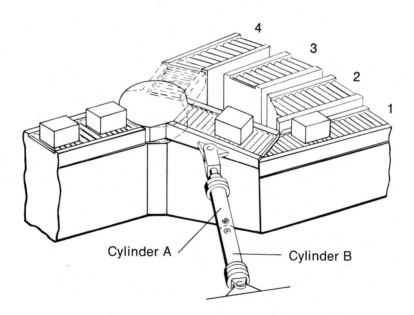

2.8.1.1 Solution for exercise 8.1

Recording of conditions in tabular form:

Position	Cylinder A	Cylinder B
1	0	0
2	1	0
3	0	1
4	1	1

Fig. 2/144 shows the circuit diagram developed from this listing.

Fig. 2/144

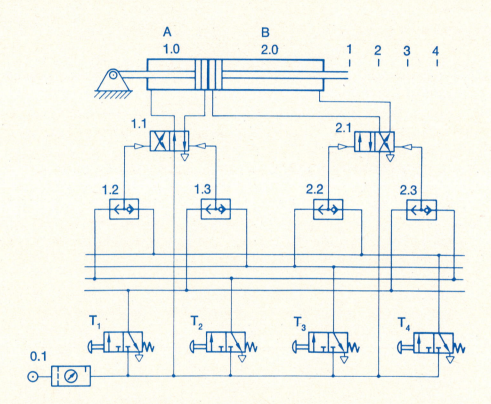

2.9 Constructing the circuit diagram for time-schedule controls

Here, it is in the first instance necessary to make the correct selection of program timer.

The determining criteria for selection include number of output instructions, length of program, frequency of program change, processing times. For pneumatic controls, the following may be considered:

- cam plates
- cam discs
- program drums
- program belts
- punched cards
- punched tapes

When designing and planning time-schedule controls, difficulties occur mainly because the speeds of the working elements cannot be precisely defined beforehand. Consequently, precise timing cannot be readily determined and overlapping of motions can occur.

With critical program parts, it must therefore be ensured that certain time safety zones are incorporated and that overlaps resulting from these are avoided. This does however involve in most cases an increase in the total running time of the program.

Since the design of a time-schedule control depends primarily on the selection of the program transmitter and no generally valid rules can be made for constructing the circuit diagram, an example will not be given at this point.

Of utmost importance, and this is again emphasized, is the exact determination of the timing of the sequence of operations and allowance for any overlaps. The sequence of motions is best represented here by the displacement-time diagram. If this diagram has been prepared, programming can be commenced, allowance being made for the program transmitter being used.

For example, if a cylinder B is to travel out when cylinder A has reached the forward end position, a displacement-time diagram for this operation will resemble that shown in Fig. 2/145.

Fig. 2/145

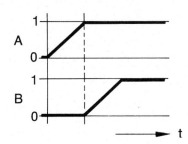

For the time-schedule type of control, this means that an overlap of the cylinder motions occurs if the outward speed of cylinder A increases. If this is to be avoided to a large extent, the safety zone mentioned above must be included between the end of motion 1 and the start of motion 2, and the magnitude of this zone is dependent on the range of variation of the advance speed of cylinder A.

In Fig. 2/146, the safety zone is represented by the time interval Δt.

Fig. 2/146

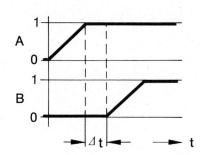

It must also be noted that when the system has been installed the specified times must be accurately set and maintained, or alternatively that the specified speed tolerances may not be exceeded, as otherwise the operation of the whole control can be disturbed.

Part 3: Control Problems, Examples and Solutions

This section contains examples selected at random of controls which have been applied in practice. The circuits shown represent **one** of the possible solutions.

3.1 Clamping fixture

Problem:

Parts are to be clamped by means of a pneumatic cylinder; initially the part is to be held by a low force to allow it to be adjusted (hand-operated push button) and subsequently it is to be clamped with full force (foot-operated push button). When the operation has been completed, the cylinder is reversed by means of another hand-operated push button.

Positional sketch:

Fig. 3/1

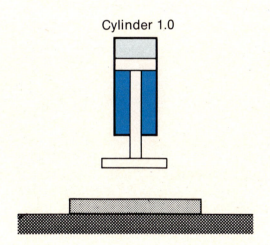

3.1.1 Circuit for exercise 3.1

Fig. 3/2

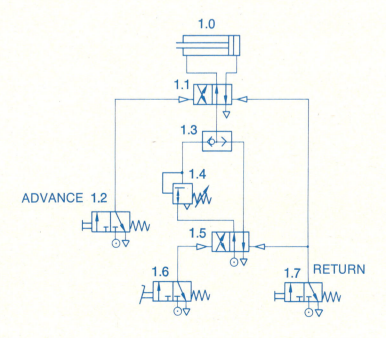

3.2 Door control

Problem:

Two sliding doors are to be controlled by two cylinders such that they can be opened from inside or from outside by operating a push button and such that they can be closed again when these buttons are operated a second time.

Positional sketch:

Fig. 3/3

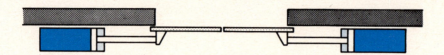

Sequence diagram:

Fig. 3/4

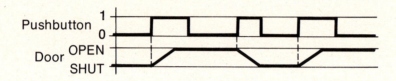

3.2.1 Circuit for exercise 3.2

The design shown in the circut in Fig. 3/5 represents one of the possible solutions. The essential part of this is the alternating circuit described in section 2.4.8.1.3 in which the output signal is alternately switched on and off or alternated.

Fig. 3/5

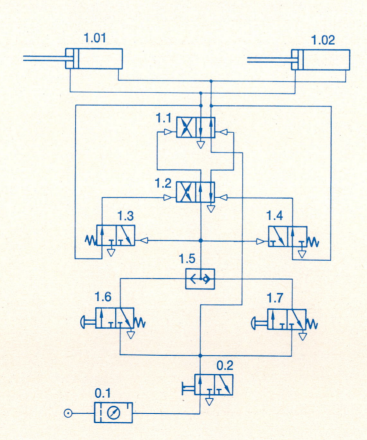

© by FESTO DIDACTIC

144

3.3 Bending fixture for spectacle frames

Problem:

Spectacle frames are bent in the middle on an automatic machine. The parts are taken from a gravity-feed magazine and pushed into the two working positions by a multi-position cylinder. First, the frame is warmed in position 1 by a tool which is pushed down by a cylinder and then bent in position 2 by a bending die operated by a cylinder. With both operations, i. e. warming and bending, it must be possible to obtain a certain adjustable dwell time in the respective end position with the tools. The forward end positions of these cylinders may not be sensed by means of limit switches. Ejection of the bent parts takes place mechanically when the positioning cylinders return.

Auxiliary conditions:

- Choice between continuous operation or single cycle.
- Magazine sensing
  Here, the system is to remain stationary in the starting position if the magazine is empty.
  Because the magazine cannot be monitored directly, the bending cylinder must be used as signal transmitter. If there are no more parts, a signal is provided from a limit switch when this cylinder travels out further.

Positional sketch:

Fig. 3/6

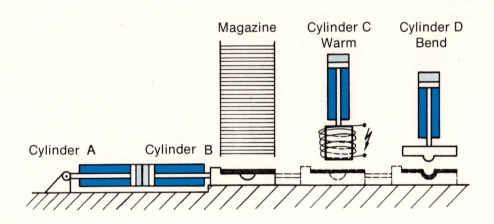

Displacement-step diagram:

Fig. 3/7

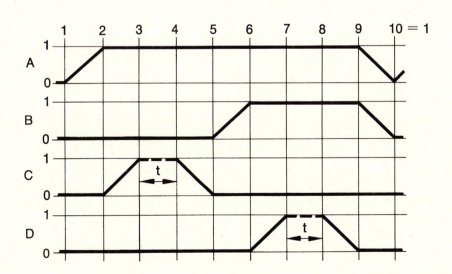

3.3.1 Circuit for exercise 3.3

Fig. 3/8

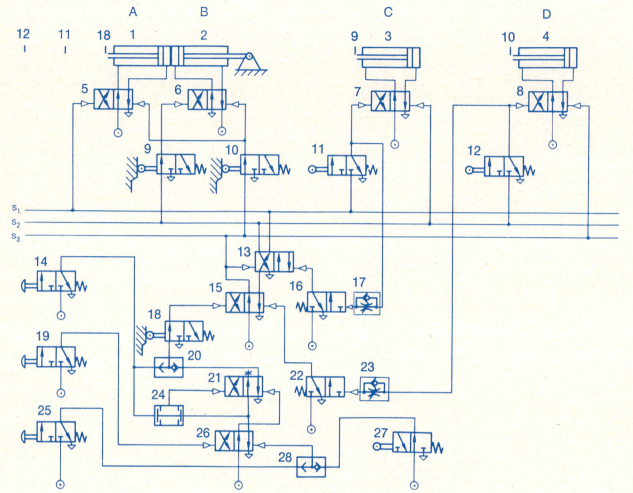

3.4 Elevator control

Problem definition and conditions:

A pneumatic freight elevator is to transport goods from the 1st floor to the 2nd floor. The elevator is controlled from the outside, either from the upper or from the lower level. The up or down signals may be effective however only if the elevator is in one of the final positions and both doors are closed. In addition, the doors are to be secured by a locking cylinder such that they can open only when the respective final position has been reached. If there is a power failure, both doors should be unlocked and the elevator should be blocked at the upper floor by means of a further cylinder in the event that the elevator is at this level at the time of power failure.

Positional sketch:

Fig. 3/9

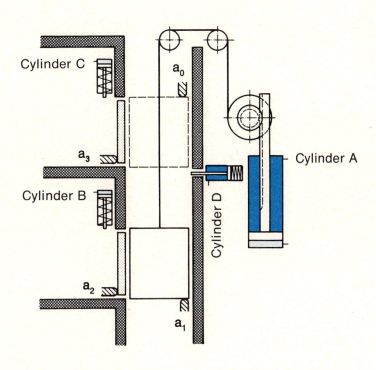

Designation of elements:

- A: Working cylinder for elevator movement
- B: Locking cylinder door lower level
- C: Locking cylinder door upper level
- D: Retaining cylinder in event of power failure

- a_0: Sensing of elevator final position top
- a_1: Sensing of elevator final position below
- a_2: Sensing of door position below
- a_3: Sensing of door position top

3.4.1 Circuit for exercise 3.4

Fig. 3/10

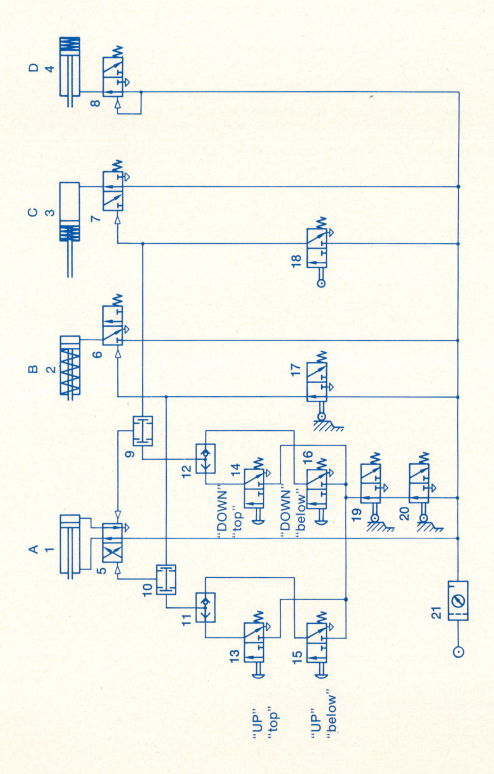

3.5 Transporting of section material

Problem:

Various plastic sections are to be transported between two caterpillar belts. The distance between the belts and the pressing force are adjusted by means of a pneumatic cylinder controlled by two push buttons.

Requirements:

— Infinitely variable and uniform opening and closing of the belts by single button operation.

— Smooth contacting without bounce on the section.

— In the event of danger, immediate opening is required by operating an EMERGENCY STOP button.

Operating speed: approx. 50–80 mm/sec.

Positional sketch:

Fig. 3/11

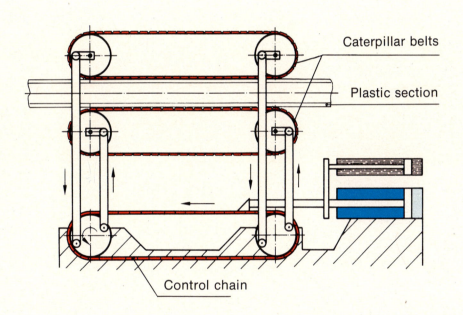

3.5.1 Circuit for exercise 3.5

Fig. 3/12

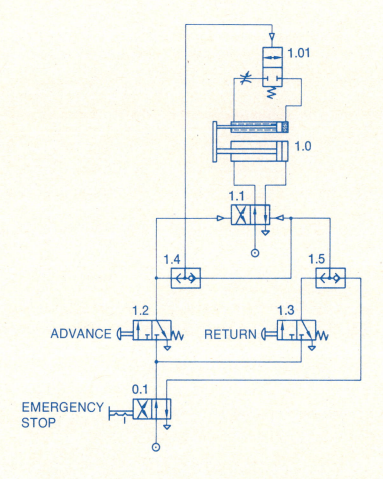

As far as at all possible by means of pneumatics, requirements 1 and 2 can be satisfied by the hydraulic check cylinder which is connected in parallel to the pneumatic cylinder. The relatively low speed allows for the most part operation without jolts. Furthermore, very high stability is obtained in the intermediate positions by the oil circuit connected in parallel.

The emergency-stop valve 0.1 ensures that the cylinder travels back to its starting position from any other position as soon as this valve is operated.

Part 4: Appendix

4.1 Examples of circuit diagram drawings

Circuit diagrams should be drawn in accordance with the standards specified in section 2.3. Two complete sample circuit diagrams are included in this appendix. The following guidelines should be observed:

Summary

- Circuit diagram layout as for control chain flowchart, signal flow from bottom to top wherever possible.
- Energy supply from bottom to top. Can be shown in simplified form.
- Spatial arrangement of the elements is ignored. Draw cylinders and directional control valves horizontally wherever possible.
- Use the same designations to identify the same elements in the circuit diagram and in the completed installation.
- Identify the location of the signals by a mark. If signals are issued in one direction only, arrow on the mark.
- Show devices in the initial control position. Identify operated elements by a mark.
- Graphic symbols for the elements in the circuit diagram to be the same as on the devices in the completed installation.
- Where possible, draw pipelines in straight lines without crossovers.
- If required, also record pipeline designations.
- If required, enter technical data, setting values etc.
- Enter sequence of movements, auxiliary conditions, and parts list.

Fig. 4/1

Parts list

Item	Name	Quantity per item	total	Still in position	Ref. No.	Comments